T0292991

Iggy Bassi and Karan Chopra, two accomplished and highly respected leaders within the world of sustainability, have written a practical climate action guide for leaders and stakeholders in every industry. More than a definitive overview of the climate crisis, this book is an economics and strategy-based road map for leaders who recognize that their climate-related actions can result in notable financial growth and reduced risks in their organization. It's a must-read for empowering data-driven, positive environmental change.

—**Dr. Jonathan Reichental,** Founder & CEO of Human Future, Professor, and Author

Decision makers who really aim to contribute to build a sustainable and more just society have to take climate science into account, and push to achieve net zero greenhouse gas emissions as quickly as possible. This book offers a valuable guide on how to achieve this for business and policy leaders.

—**Professor Roberto Buizza** (*Physics*), Interdisciplinary Centre on Sustainability and Climate, Sant'Anna School of Advanced Studies, Italy, former Lead Scientist at ECMWF

As organizations lean into a constellation of climate and nature related risks and opportunities, the concept of Unified Climate Intelligence (UCI) can enable the creation of environmental, social and economic value. This book offers important guidance for organizations wanting to harness UCI to lead in the climate economy.

—**Munish Datta,** Director of Sustainability, Fellow, Advisor, Trustee

Iggy and Karan have developed the essential operating system for CEOs, leadership teams, board members and investors to harness future intelligence in tackling the climate crisis. Their Unified Climate Intelligence empowers resilient, climate-aligned decisions and actions, enabling systemic adaptation over superficial "fixes".

—**Roger Spitz,** President, Techistential and Chair, Disruptive Futures Institute

Praise for
The Climate Intelligent Organization

It is exciting to see such a practical approach to addressing climate change, seeking opportunities to improve rather than being despondently mired in crisis. Climate change has become a social movement as much as physical changes, so businesses and governments need UCI's guidance to stay atop the rapidly shifting political and environmental landscapes.

—**Prof. Ilan Kelman,**
University College London

If you care about future-proofing your business for climate vulnerabilities, this book is essential reading. A thorough and insightful guide for leaders to build differentiated products and create value in an ever-changing world.

—**Vrushali Gaud,** Director at Google and
former Managing Director of Climate and
Sustainability Services at Accenture

Doing what they do best, climate data intelligence, Iggy and Karan provide leaders with a blueprint for value creation amidst climate change.

—**Courtney Elizabeth Stipe Holm,** The Holm Edit and
Holm Advisory (former Global VP Impact,
Sustainable Futures at Capgemini)

Sure-footed and admirably clear, this is an invaluable guide to navigating the greatest challenge of our time.

—**David Shukman,** Speaker, Consultant and
former BBC News Science Editor

We hear a lot about the state of climate change, but not nearly enough about the practical steps businesses can take to capitalize on digital technologies and truly integrate climate thinking in their operations. I am excited to see how Unified Climate Intelligence changes the game, hopefully speeding up the energy transition and reducing its inevitable bumpiness.

—**Diana Fox Carney,** Advisor, Helios
Climate and Eurasia Group

As Darwin found, it's not the fittest that survive in a changing environment, but the most adaptable. Based on their deep understanding of the latest technology and a rich experience as entrepreneurs, Iggy Bassi and Karan Chopra offer tools and insights for organizations not just to survive, but to thrive. A veritable treasure trove for business leaders looking for new, positive cohesive strategies for opportunity and success in a changing climate reality.

—**Maurits Dolmans,** Senior Counsel,
Cleary Gottlieb Steen & Hamilton LLP,
London. Advisor to UNFCCC

Too many businesses are flying blind into an unpredictable future, mistaking their rear-view mirror for their steering wheel, missing essential signals that are hiding in plain sight. This book will show you how to sift the signals from the noise, and navigate an endlessly shifting landscape of the climate challenge and the meta-crisis world.

—**Gil Friend,** CEO, Natural Logic, Inc., MD,
Critical Path Capital, and former Chief
Sustainability Officer of the City of Palo Alto

THE CLIMATE INTELLIGENT ORGANIZATION

THE CLIMATE INTELLIGENT ORGANIZATION

BUILD A PROSPEROUS AND RESILIENT FUTURE FOR THE PLANET THROUGH AI-POWERED CLIMATE INTELLIGENCE

IGGY BASSI
KARAN CHOPRA

WILEY

This edition first published 2025

Registered Office(s)
John Wiley & Sons, Inc., 111 River Street, Hoboken, NJ 07030, USA
John Wiley & Sons Ltd, The Atrium, Southern Gate, Chichester, West Sussex, PO19 8SQ, UK

Editorial Office
The Atrium, Southern Gate, Chichester, West Sussex, PO19 8SQ, UK
For details of our global editorial offices, customer services, and more information about Wiley products

visit us at www.wiley.com.

Library of Congress Cataloging-in-Publication Data is Available:

ISBN 9781394192397 (Cloth)
ISBN 9781394228553 (ePDF)
ISBN 9781394228539 (ePub)

Cover Design: Wiley
Cover Image: © Floriana/Getty Images

SKY10092982_120424

This book is dedicated to our families, whose patience and unwavering support have been our anchors through the challenges of pioneering climate solutions. A special mention to our children – Jasmin, Camilla, Surina, and Kalil – who inspire us daily to address the climate crisis with urgency and hope. We also extend our deep gratitude to the climate scientists and researchers, whose groundbreaking work forms the bedrock of climate intelligence. To the visionaries and entrepreneurs driving the climate movement, and the forward-thinking companies and their internal champions: your bold leadership is crafting a world that harmonizes human activity with the rhythms of nature. Your collective endeavors are essential to safeguarding the human project.

CONTENTS

PREFACE AND INTRODUCTION

A PERSONAL JOURNEY THROUGH THE CLIMATE DIALOGUE

Welcome to *The Climate Intelligent Organization*. This book reflects our collective experience over the past quarter century, offering a deep dive into the evolving narrative of climate change. Enriched with our insights, we aim to transform how organizations perceive and engage with the climate challenge by introducing Unified Climate Intelligence (UCI). UCI represents the next frontier in climate strategy and technology, shifting the focus from risk to creating opportunities and accelerating climate-aligned growth with tangible value creation.

Our journey began in 2008, while advising a sovereign wealth fund manager on emerging economies. Dismissed then as irrelevant, climate change is now at the forefront of economic strategy for that same entity. This dramatic shift underpins the global awakening to the urgency of climate action – a transformation we have been a part of, advising companies and governments on competitiveness, growth and resource security. Our professional paths have taken us from observing sustainability and climate change as mere afterthoughts to driving the narrative toward urgent, actionable change, where opportunities for innovation, value, and risk management coexist. This personal journey, filled with both frustrations and triumphs, is the wellspring from which *The Climate Intelligent Organization* originates.

Our experiences have revealed a stark disconnection between environmental sustainability and economic decision making. This realization spurred a profound shift in our careers. We ventured into sustainable agribusiness in West Africa, confronting first-hand the harsh realities of climatic volatility on food security. This venture was not just about business; it was about deeply understanding and mitigating the impacts of climate change at a community level. The resilience and adaptation strategies we developed there were early iterations of what we now define as UCI.

However, our pivotal moment came with the founding of Cervest, a data analytics platform designed to predict and quantify physical climate hazards. Launched as nations around the world committed to the Paris Agreement, Cervest and its flagship product EarthScan was our response to the critical need for tools that could translate complex climate data into actionable business insights. We quickly recognized that complex data, when transformed into decision-useful insights, could help companies not only understand but also communicate climatic impacts to regulators demanding new answers. Our frustration grew as climate discussions remained peripheral, with no binding agreement until the Paris breakthrough, yet climate risks and CO_2 levels continued to rise. Our first-hand experience on the farm, especially when a "50-year wind event" damaged half our grain milling factory, taught us the urgent need for predictive climate intelligence. If we had possessed better insights earlier, we would have altered the engineering decisions for our food processing factory and farming infrastructure to better withstand such events.

From Risk to Opportunity: A Quantum Leap with Unified Climate Intelligence

For too long, climate change has been viewed predominantly as a paralyzing risk, overshadowing its potential as a catalyst for innovation and value creation. Our professional journeys, characterized by both challenges and insights, have illuminated the limitations of viewing climate solely through a lens of risk. This perspective not only stifles the urgency for proactive measures but also blinds us to the vast opportunities that proactive climate action can present.

However, the narrative is changing. A technological revolution is well underway, reshaping our approach to climate challenges. Advancements in AI, machine learning, big data and the declining costs of clean technologies are ushering in a new era. These developments allow us strategic foresight to understand and react to climate dynamics much earlier and more accurately than ever before, unlocking tremendous value for businesses and society alike.

The emergence of radical transparency in climate reporting fundamentally changes the business landscape. With increased visibility into climate risks and opportunities, there is no longer "anywhere to hide." This transparency compels businesses to be proactive in managing their climate impact, not just for compliance, but as a strategic imperative to maintain a competitive advantage. Previously, a company's climate strategy might have been a private matter. Now, these strategies are openly quantified and scrutinized by B2B partners, banks, regulators and insurance companies. This introduces a game theory dimension to climate action. Furthermore, advancements in Generative AI (Gen-AI) hold immense potential to revolutionize and democratize climate intelligence. Gen-AI can be harnessed to analyze vast datasets, identify complex patterns, generate more nuanced climate insights and make this all contextually relevant in natural language. This, in turn, can inform the development of even more sophisticated UCI frameworks, enabling businesses to permanently transform their relationship with climate by making data-driven decisions that not only mitigate risks but also capitalize on emerging opportunities in the low-carbon economy.

Companies with robust climate strategies that demonstrably mitigate risk and capitalize on opportunities will be seen as more attractive partners, more creditworthy borrowers, and more insurable entities. Conversely, those lagging behind will face disadvantages in all these areas.

UCI is at the forefront of this revolution, transforming how we analyze climate data. Traditional methods, often fragmented and isolated, fail to provide a holistic view of climate impacts. UCI transcends these limitations by integrating diverse data sources – ranging from climate science and financial metrics to AI analysis – into a cohesive framework. This comprehensive approach provides a 360° view of climate risks and opportunities, empowering businesses to:

- Identify and Prioritize Climate Action: Gain granular, quantified insights into how climate change and related policies could impact operations, supply chains and market dynamics.
- Discover Hidden Opportunities: Explore new markets catalyzed by the low-carbon transition, including innovative clean technologies.
- Make Data-Driven Decisions: Embed climate considerations into core business strategies to enhance decision making and future-proof operations.
- Enhance Resilience: Develop robust strategies to adapt to the changing climate and minimize potential disruptions.
- Reduce Costs: Achieve greater resource efficiency and energy savings, enhancing operational efficiency and profitability.

By embracing UCI, businesses can shift from a risk-averse posture to one where climate action is a strategic driver of innovation, value creation and long-term resilience. This shift is not just theoretical but practical, as we have moved from focusing solely on risk to leveraging climate action as an opportunity for profound economic and social transformation. We see immense potential for climate intelligence to accelerate the transition toward climate-aligned enterprises and cities, enhancing their viability, livability and economic prosperity.

This holistic understanding of UCI, which incorporates physical hazards, transition pathways, CO_2 removal, nature conservation, resource optimization and fiscal incentives, is crucial. It allows for a deeper comprehension and management of climate-related financial performance, propelling businesses and cities toward actions that are not only environmentally sustainable but also economically beneficial and socially necessary.

Beyond CO_2: A Multifaceted Approach for a Polycrisis

The world has gone all-in on net zero since the Paris Agreement, aiming to achieve a future where human activity no longer adds greenhouse gases to the atmosphere. This ambitious goal is undeniably necessary – it's the cornerstone of mitigating climate change and preventing catastrophic warming. However, a singular focus on net zero is myopic. We are living in a time of multiple, interconnected crises – a polycrisis. Climate change, biodiversity loss and ecological degradation are intricately linked, forming a complex web of environmental challenges.

The current approach, where many companies set far-off net zero targets and often rely on carbon offsets to achieve them, is a dangerous oversimplification. While carbon offsets play a role, their limitations are becoming increasingly apparent. For instance, recent controversies surrounding the sourcing of carbon credits, such as those linked to forest fires in BP and Microsoft's case, highlight the need for permanence and robust verification systems. Focusing solely on decarbonization goals neglects the interconnected nature of these crises. As Iggy conveyed to the former US vice president Al Gore at COP 26, "we simply can't decarbonize our way to a climate-aligned economy." Reaching net zero emissions is crucial, but it's not the only piece of the puzzle. We need a multifaceted approach that factors in physical hazards, the depletion of natural capital and strategies for carbon removal.

The consequences of this short-sighted approach are already unfolding. Climate hazards are exceeding the engineering design and resource capacity of many cities and infrastructure systems. Rising sea levels, extreme weather events, and heat waves are pushing these systems to their breaking points. Understanding the lag effect inherent in the Earth's systems is crucial it's akin to turning off a boiling pot of water; although we stop the heat in 2030, the water will continue to bubble for a long time. This delayed response means that even if we achieve net zero emissions tomorrow, the phenomena we've set in motion – like rising sea levels due to thermal expansion of oceans and melting glaciers, ocean acidification from increased CO_2 absorption and ongoing global warming due to the current energy imbalance – will persist for decades. This enduring impact underscores the need for robust adaptation strategies, particularly to mitigate the devastating effects of heat. We must now prepare for a permanently altered climate, a change that will span generations.

Leadership and the Climate-Intelligent Journey: A Tailored Approach

Strong leadership is the cornerstone of successful climate action. We've consistently observed that companies, cities, and countries that are ahead of the curve – those demonstrating progress in building a more resilient and equitable future – all have clear and decisive leadership at the helm. These leaders understand the urgency of climate action and possess the vision and commitment to translate that understanding into concrete strategies that drive value. Understanding and managing your climate-related financial performance will become a core competency as we transition to a climate economy.

The Climate Intelligent Organization serves as a road map for leaders across all sectors, empowering them to embark on their unique "climate journey." This journey is not a one-size-fits-all proposition. Every organization, regardless of size, industry or location, faces a distinct set of challenges and opportunities when it comes to climate action. However, the tools we now possess, like UCI, can significantly accelerate progress on this journey.

The Role of the Climate-Intelligent Leader

Climate-intelligent leaders go beyond simply recognizing the importance of climate action. They actively champion it, fostering a culture of environmental responsibility within their organizations. Here are some key characteristics that define them:

- Vision and Strategic Thinking: They possess a clear vision for a climate-resilient future and translate that vision into quantified value and actionable strategies, aligned with the organization's overall goals.
- Decision Making with Climate in Mind: Climate considerations are integrated into all aspects of decision making, from resource allocation to product development and investment strategies.
- Communication and Transparency: They effectively communicate the organization's climate strategy and its rationale to stakeholders, fostering transparency and buy-in.
- Collaboration and Partnerships: Recognizing the interconnectedness of the climate crisis, they actively seek out partnerships with other organizations and stakeholders to drive collective change.

Tailoring Climate Intelligence: A Framework for Action

While the core principles of climate action are universal, the specific actions taken by organizations will vary depending on their context. This book equips you with the knowledge and tools to tailor climate intelligence to your organization's specific needs, offering a framework to navigate your unique climate journey effectively.

- Climate Risk and Opportunity Assessment: The first step is to understand your organization's unique vulnerability to climate hazards, as well as potential opportunities that arise from transitioning to a low-carbon economy. Tools like UCI are invaluable in this assessment, helping to identify where the future performance of your business could be most impacted or could benefit from innovative changes. UCI helps you translate climate risks and opportunities into financial terms, enabling you to understand your climate related financial performance in the past, present and future.
- Setting Ambitious Yet Achievable Goals: Once you understand your baseline, it's crucial to establish ambitious yet achievable climate goals

that align with your organization's long-term vision. These goals could range from reducing emissions to enhancing resource efficiency, or even devising adaptation strategies to tackle specific climate risks. UCI empowers you to prioritize the areas of focus, to understand the cost of inaction and to quantify financial value that can be unlocked by achieving your goals. The aim is not just to mitigate risks but to also seize opportunities that drive competitive advantage and value creation.

- Developing a Climate-Intelligent Strategy: Translate your goals into a comprehensive strategy that outlines the specific actions your organization will undertake. This strategy should encompass operational changes, technological advancements, stakeholder engagement and an economic reevaluation to ensure sustainability efforts align with profitability. In the climate age, value creation and competitive dynamics must sync with environmental strategies, providing a robust framework for growth and resilience.
- Implementation and Monitoring: Effective implementation of your climate strategy is crucial. Allocate resources wisely, establish clear ownership of initiatives and set up robust monitoring and evaluation systems to track progress and impact. This phase is vital for adjusting strategies in real-time and demonstrating tangible benefits both internally and in the wider market.
- Continuous Improvement: The climate landscape is constantly evolving, and so must your strategies. Regularly review your progress and adapt your approach as needed, staying informed about the latest advancements in climate intelligence tools and technologies. As risks intensify, there will be a shift toward policies that focus on de-risking through value-based approaches, influencing B2B relationships, trade agreements, and market access.
- By following this framework and leveraging the transformative power of UCI, leaders across all sectors can embark on their unique climate journeys. This approach not only builds climate-intelligent organizations but also positions them advantageously within new competitive equations that consider environmental impact as a core element of business success.

A Stark Reality: A Global Crisis Demands a Global Response

Climate change is the defining challenge of our generation. Since the landmark Earth Summit in Rio de Janeiro in 1992, a turning point in global environmental awareness, greenhouse gas emissions have continued to rise relentlessly. Despite international agreements and a growing public outcry, meaningful progress remains elusive. Time is a luxury we can no longer afford.

This book is not intended to be a chronicle of despair, but a clarion call to action. The science is unequivocal - the longer we delay decisive action, the more severe the consequences will be. The impacts of climate change are no longer hypothetical; they manifest globally through extreme weather events, rising sea levels and ecological disruptions. Twenty years ago, climate change was dismissed as primarily a problem for developing economies. This view grossly underestimated the interconnectedness of our planet. The past two decades have provided a harsh lesson as extreme weather events, rising temperatures and environmental degradation have become relentless realities impacting every corner of the globe - developed and developing economies alike. Climate change transcends borders and economic systems; it demands a global response.

In recent times, the acute risks posed by the COVID-19 pandemic demonstrated how rapidly global entities can mobilize in the face of imminent threats. Climate change, while often perceived as a "disaster in slow motion," requires the same level of urgency and unified action. Since the 1992 Earth Summit, CO_2 levels have continued to rise, only dipping briefly during the global lockdowns of COVID-19. We are currently on track for potentially catastrophic climate changes, with projected global temperature increases of up to 2.7°C, carrying severe consequences.

Yet, we are not without hope. The tools, technologies and ingenuity at our disposal today are unprecedented, enabling us to make significant advancements. This new arsenal of climate intelligence and technologies are galvanizing change, particularly among the younger generations who stand to inherit an Earth on the precipice, with the potential to be more liability than asset. These young people, globally connected and soon to be the world's primary economic drivers, demand a sustainable and equitable world.

Building a Climate Intelligent Future: A Roadmap

This book is carefully structured to provide you with the essential knowledge and tools needed to integrate climate intelligence into your organizational strategy. Part I: The Imperatives of Climate delves into the scientific foundations and the critical challenges and opportunities presented by climate change. This section explores the multifaceted nature of climate change, encompassing its impact on various sectors, alongside the evolving landscape of global climate policy. A dedicated chapter dives deeper into the intricate interplay between national and international policy frameworks, city-level initiatives and sectoral efforts in driving climate action. This comprehensive understanding is crucial for businesses to navigate the evolving regulatory environment and identify opportunities for alignment with climate-positive policies. This section sets the stage by laying down the foundational concepts necessary to grasp and implement UCI effectively within any organizational context.

In Part II: Building a Climate Intelligent Organization, the focus shifts from theoretical understanding to practical application. It delves into how businesses can harness UCI to drive strategic decision making and secure a competitive advantage. This segment illuminates how advancements in artificial intelligence, climate science and analytics have revolutionized our approach to managing climate data, offering businesses unprecedented insights into climate risks and opportunities. We explore the integration of UCI into corporate strategies, demonstrating how forward-thinking companies have successfully navigated the transition to a climate-conscious economy. This section underscores the importance of radical transparency in business operations and sustainability in corporate governance, providing real-world examples and case studies that showcase the potential for businesses to not only comply with emerging regulations but also seize growth and innovation opportunities in an environmentally-aware market landscape.

This book is not merely a theoretical exploration; it is a practical guide tailored for business practitioners, policy executives and public sector leaders. It aims to redefine the narrative around climate change from one of risk to one of opportunity and value creation. It also sheds light on the tools and technologies at our disposal to empower leaders to make climate aligned decisions that benefit the health of their businesses. Through the lens of Unified Climate Intelligence, decision makers across enterprises, cities, states and non-profits will be equipped to effectively prioritize resources, mitigate risks, adapt and foster sustainable development and new sources of value.

As we embark on this journey together, we invite you to join us in unlocking the immense value of climate action. By following the structured approach outlined in this book, your organization can transform its business model for a sustainable future, building a more equitable and resilient world for the next generation. Together, let's catalyze societal change and foster transformative leadership within the climate age.

PART I
The Imperatives of Climate

CHAPTER 1
THE GREAT ACCUMULATION PROBLEM

Summary

In this chapter, we explore the profound impact of climate change, something that is no longer a distant, debatable concept but a global, scientifically irrefutable reality. Climate risks, irrespective of financial status or geographic location, affect everyone. However, vulnerable communities, assets and physical structures, often in developing nations, bear the brunt of these impacts due to factors like geographical location and socio-economic conditions and engineering design.

The central driver of global warming is the accumulation of greenhouse gases (GHGs) in the atmosphere. Carbon dioxide (CO_2) and methane (CH_4) are pivotal GHGs due to their direct emissions from human activities. CO_2 levels have surged since the Industrial Revolution, exceeding natural variability by far. Additionally, human activities impede natural processes that sequester CO_2. Methane, though less abundant, is much more effective at trapping heat. Its levels have risen significantly, driven by both natural and human-induced sources.

Global warming intensifies climate risks, amplifying the frequency, severity and duration of extreme events, sea-level rise, and ecosystem disruptions. Climate risk encompasses physical and nature risks and transition responses, all of which are interconnected and demand comprehensive solutions. Population growth, coupled with rising lifestyle expectations, strains the planet's resources and contributes to GHG emissions. However, addressing these issues through sustainable practices, education and awareness is essential. Climate risk is not distributed evenly, with developing nations facing greater vulnerability.

Tackling climate risk is complex, as it transcends borders and spans long timeframes. Compounding climate risk, arising from the interaction of climate change, existing vulnerabilities and regulatory and economic factors, adds to the challenge. Urgent action is imperative, as the costs of climate-related damage are already substantial and projected to escalate without intervention. International cooperation and private sector involvement are pivotal in addressing climate risk effectively. The chapter underscores that success in this endeavour is a collective win or loss for humanity.

Climate change is no longer only in the realm of science; it has become part of all of our vocabulary as the impacts become tangible across the globe. The increasing risks from climate change are non-discriminatory, affecting lives and livelihoods worldwide regardless of financial wealth, social structure or hemispheric location. However, the impacts of this more volatile climate are not experienced equally and are closely linked to preparedness; the resilience to and ability to recover from climate change-related events. Vulnerability to climate impacts depends on various factors, including geographical location, socio-economic conditions and adaptive capacity. As such, climate change is inherently more damaging to communities in developing countries and assets that have not been engineered to withstand such fluctuations.

However, the common denominator across all climate risk and impacts is rising global temperature. Earth's average global temperature has warmed by 1.2°C since 1850 – a significant point in time due to its association with the Industrial Revolution. At the time of writing, the last decade has been the warmest ever recorded and the most costly in global economic losses. While an average rise of 1.2°C globally may sound innocuous, the resulting extremes in temperature change are felt much more severely on a regional scale with heat waves now often exceeding 40°C in continental Europe, and 50°C in parts of Africa, Asia and the Middle East. These once unusual temperature extremes have moved far beyond what many species, societies and businesses have lived through, and are now estimated to be 100 times more likely to occur as a result of climate change (Fraser-Baxter 2023). More volatile climate events are becoming the norm; hurricanes are becoming stronger, droughts are lasting longer and wildfires are spreading faster. Rising sea levels are affecting coastal areas, posing a threat to cities and communities along coastlines.

Outside of direct weather volatility experienced by an increasing percentage of our populations, changing weather patterns make it harder for farmers to predict when to plant and harvest crops, disrupting food chain supply and causing price fluctuations. Health risks are increasing, with more people facing heat-related illnesses, and diseases are spreading to new areas as temperatures shift entire ecosystems. Economies are being impacted, too. Industries like tourism and insurance face growing risks. Businesses are having to adapt to disruptions in their supply chains and higher costs for resources.

The increase in global average temperatures has been driven by human activity – specifically the release of greenhouse gases (GHGs) into the atmosphere. We will dive deeper into the science behind climate change in the next chapter; first let's explore the phenomenon of escalating atmospheric GHG – The Great Accumulation Problem.

GHGs are like a natural blanket in our atmosphere. They allow sunlight to enter the Earth's atmosphere, but they also trap some of the heat that tries to leave playing a crucial role in regulating the planet's temperature and making it suitable for life as we know it. However, when we burn fossil fuels such as coal, oil and gas for energy and make changes to the land such as deforestation, we release extra GHGs. This thickens the natural blanket, leading to more heat being trapped and ultimately to global warming.

The most commonly discussed GHGs are carbon dioxide (CO_2) and methane (CH_4). While these are not the most abundant GHGs in Earth's atmosphere, the focus remains on them due to their direct emission from human activities and their significant contributions to the enhanced greenhouse effect. This is also the reason that many measurements, models, and emissions scenarios use "pre-industrial" CO_2 concentrations as a baseline to compare current levels; as a marker for conditions before significant human contributions.

Carbon dioxide levels have increased at an unprecedented rate since the late eighteenth century. Prior to the Industrial Revolution, CO_2 levels were relatively stable at around 280 parts per million (ppm). This stability had been maintained for thousands of years. However, with the advent of industrialization, there was a rapid increase in the burning of fossil fuels for energy, as well as increased deforestation. This led to the initial release of large amounts of CO_2 into the atmosphere, followed by a more gradual increase during the mid-twentieth century. Into the second half of the twentieth century, CO_2 concentrations started to increase at an accelerated rate due to the widespread adoption of fossil fuel-based technologies, particularly in transportation, industry and energy production. By the end of the twentieth century, CO_2 levels had reached 370 ppm, well above the natural variability captured in ice core records covering the past 800,000 years. In the past couple of decades, the accumulation of CO_2 has been unprecedented, exceeding 400 ppm and largely due to the expanding use of fossil fuels in developing economies, as well as ongoing industrialization and urbanization in many parts of the world.

As well as emitting CO_2 at a more rapid rate and increasingly accumulating atmospheric concentrations, human activities are also responsible for reducing many of the natural mechanisms that remove – or sequester – CO_2 from the atmosphere naturally such as forests and peatland. This adds to the accumulation problem by making it more challenging for natural systems to offset emissions from human activity; a topic that we will explore further in the next chapter.

Methane is less abundant in the atmosphere compared to CO_2; however, it is nearly 30 times more effective at trapping heat in the atmosphere over a 100-year period, making it a significant concern to mitigation practices. Methane is released into the atmosphere from both natural processes and human activities. Natural sources include wetlands, oceans and gas hydrates. Human-made, often referred to as "anthropogenic" sources include agriculture, particularly livestock digestion and rice cultivation, fossil fuel production and use and certain industrial processes. Concentrations have significantly increased since the late eighteenth century and are showing a rapid uptick in the most recent decade. Potential sources for this include increased emissions from agriculture, changes in fossil fuel production and distribution and releases from natural sources due to environmental changes as global warming continues. Methane has a much shorter atmospheric lifespan of around 12 years compared to centuries for CO_2, and so while it is a substantial driver of near-term global warming, focused efforts to remove CH_4 alongside the reduction of further emissions also have the unique power to yield relatively rapid benefits in terms of slowing down the pace of global temperature rise. By "buying time" through these actions, CH_4 reduction complements broader climate change mitigation strategies that are aimed at long-term stabilization of climate change.

Global warming is the direct result of GHG accumulation in the atmosphere. Elevated temperatures contribute directly to climate risk by disrupting the Earth's major systems, ultimately increasing volatility and unpredictability of climate by amplifying the duration, frequency and severity of extreme weather events, rising sea levels and disruptions to ecosystems. Combined, these risks pose significant challenges to communities, economies and natural systems worldwide.

Climate risk encompasses several interconnected components including physical risk, natural capital degradation and biodiversity loss alongside the costs of the transition response. We will explore the concept of climate risk as well as potential opportunity arising from climate change further in Chapters 3 and 4, but here we will begin with a baseline overview. Physical risks stem from the direct impacts of climate-related events and hazards, such as acute extreme weather events (e.g., hurricanes, floods, droughts), and chronic shifts such as rising sea levels and long-term temperature increases. These events can damage infrastructure, disrupt supply chains and pose threats to human health and safety and harm the biosphere. Transition risk arises from the shift toward a low-carbon economy and the efforts to mitigate climate change. They include policy and regulatory changes, technological advancements, market shifts and reputational risks. Businesses and industries heavily reliant on high-carbon activities may face financial challenges during this transition. Nature risk, including biodiversity loss and natural capital degradation, refers to the destruction of natural habitats, ecosystems and the loss of biodiversity. This degradation leads to reduced ecosystem services, such as pollination, clean water and climate regulation, which are vital for human well-being. These components of climate risk are interconnected, and addressing them requires a comprehensive and integrated approach that involves businesses, governments, communities and individuals. Understanding and effectively managing these risks is crucial for building resilience and capitalizing on the opportunities that also exist within the face of a changing climate.

Climate risk is distinct from other types of risk due to its underlying causes, widespread impacts and the complex interplay of natural and human-induced factors. In addition to the accumulation of GHGs in the atmosphere, several other drivers exist. These include land use changes such as deforestation, urbanization or agricultural expansion, industrial and agricultural practices linked to the release of more potent GHGs such as CH_4, energy production and consumption and natural climate variability such as El Niño events, which become exacerbated when overlain on an already disrupted climate system.

The climate risk discussion must include population growth. Global population has soared from 1.2 billion in 1850, to more than 8 billion as of 2023 and with that comes consistently increasing demand for resources, energy, and food production and ultimately, rising GHG emissions. At our current size, and with accelerating lifestyle expectations, we outstrip the planet's annual sustainable resource use earlier each year. In fact, 1971 was the last year on record where we achieved even an equitable balance of supply/demand. At the time of writing, we consume a total of 1.7 Earth's worth of resources every year meaning not only do we overshoot with our demands, but we are consistently digging into our planet's resource overdraft – further contributing to environmental stresses such as water shortages. While sensitive, addressing population growth and lifestyle factors is crucial for mitigating climate risk. Sustainable urban planning, promoting energy efficiency, transitioning to renewable energy sources, encouraging sustainable diets and promoting responsible consumption are all essential strategies for reducing the impact of population growth and lifestyle choices on climate change. Importantly, education and awareness-raising play a key role in influencing sustainable behaviors and choices at an individual level and cannot be underestimated as a tool in any resilience building strategy.

As mentioned, climate risk is not distributed evenly across the globe. It varies by geography but more crucially vulnerability, meaning that developing countries are often more severely impacted. Not only is climate risk wickedly complex, characterized by interconnected feedback loops between natural systems, human activities and socio-economic factors; it is also uniquely challenging to tackle, transcending national boundaries, with potential resolutions ultimately exceeding human timescales to be realized. It represents an urgent and formidable challenge requiring global cooperation in every form.

Part of the complexity of planning for climate risk is that it does not behave linearly. As Earth's natural systems are disrupted due to global warming, we are experiencing changes in the probability of extreme events – the metaphorical loading of the climate dice to more frequently, albeit less predictably, land on a category 4 storm than a category 2 for example. Added to this is the complication of compounding climate risk. In this instance climate change exacerbates existing vulnerabilities and introduces new challenges at the same time; while alongside, regulatory and economic implications are also in play. Take the example of a coastal resort hit by a storm; the storm is more intense and coincides with higher sea levels. This increases the likelihood of significant damage to the resort. Additionally, the regulatory changes and higher insurance costs add financial strain. Customer cancellations further impact revenue, and supply chain disruptions delay necessary repairs.

Perhaps the largest peril of GHG accumulation is the perception of urgency with which it needs to be addressed. Climate change is not a future risk, it is now. The costs associated with climate change-related damage between 1970 and 2021 reached an estimated total of US $4.3 trillion in economic losses and resulted in more than two million deaths globally (World Meteorological Organization, 2023). An estimated 25% of those costs occurred between 2010 and 2019 alone. Without action, losses of as much as $23 trillion in reduced annual global economic output worldwide are predicted by 2050 under the worst climate scenarios. Again, this is inequitably distributed with some Asian nation economies projected at one-third less wealth than would otherwise be the case (Swiss Re, 2021). The costs of climate risk will continue to rise each year without timely and effective intervention to not only mitigate damage to existing assets, but to build climate resilience into every new policy, strategic and operational decision moving forward. To succeed at the scale and the pace required, this action must mirror the networked nature of the climate system itself with international cooperation at every level from communities to countries. Ultimately, either we all win, or we will all lose.

The growing awareness and recognition of global warming has spurred the beginning of global actions including international annual meetings such as the Conference of the Parties (COP) and global decarbonization targets based around frameworks built through policy-led cooperative work. While crucial advancements in the reduction of GHGs have been made through seminal treaties such as the Paris Agreement, they rely on commitments that are ultimately enforced on a national level and at the speed of a country's political will. A clear disparity exists between the urgency with which nations must be able to act, and the pace of legislative change and it has become increasingly clear that policy alone cannot be the driving force to address climate risk. With its autonomous resources and international remit, there is a crucial and rapidly accelerating role for the private sector to step in and turbocharge the change required to address climate risk in the immediate future.

References

Fraser-Baxter, S.E. (2023, 5 May). Study finds recent heatwave in Africa and Europe was fuelled by climate change. https://www.imperial.ac.uk/news/244753/study-finds-recent-heatwave-africa-europe/

Swiss Re (2021, 22 April). The economics of climate change. https://www.swissre.com/institute/research/topics-and-risk-dialogues/climate-and-natural-catastrophe-risk/expertise-publication-economics-of-climate-change.html

World Meteorological Organization. (2023, 22 May). Press release. https://wmo.int/news/media-centre/economic-costs-of-weather-related-disasters-soars-early-warnings-save-lives

CHAPTER 2
THE SCIENCE OF CLIMATE CHANGE

Contributing author: Dr. Claire Huck

Summary

In this chapter, we explore the key concepts of climate science that underpin our climate models and predictions. There is an undeniable distinction between natural and human-induced climate change, characterized by the rapid and persistent nature of current human-driven changes. While natural processes have historically regulated Earth's climate over millennia, human activities, particularly the burning of fossil fuels, have caused a significant increase in atmospheric carbon dioxide (CO_2) levels at an unprecedented rate.

The interconnectedness of Earth's major systems, including ice sheets, continents, ecosystems, atmosphere and oceans requires any response to climate change to consider these relationships and adopt a unified approach. However, we only have a limited window to remain within the 1.5°C temperature target outlined in the Paris Agreement. Exceeding this temperature will force our environment toward global climate tipping points, where irreversible consequences can be triggered such as runaway ice sheet collapse and significant sea level rise.

The role of climate models, especially Global Circulation Models (GCMs), in understanding and projecting future climate changes is crucial. These models incorporate human factors, such as greenhouse gas emissions scenarios, alongside natural environmental data and are essential for informing climate policies and international agreements, such as the Paris Agreement, and for assessing the impact of various policy measures.

At more than 4 billion years old, Earth has been through several climatic evolutions ranging from a cratered, moon-like landscape with primordial oceans, to a planet with an atmosphere containing oxygen and supporting an abundance of evolving life. Even this version of Earth, though, has experienced extreme heat, complete encasement in ice and everything in between. Hearing our modern climate change journey being described as "dangerous" or "extreme," therefore, can be a point of confusion for many. One of the most commonly asked questions in the general climate change conversation is, "but isn't this just part of natural change?" In the previous chapter, we outlined the climate basics of global warming and how this translates into climate risk; here we will take a closer look at the scientific interaction of the Earth's environmental components and the role that these interconnected systems play in responding to rising temperatures forced by human activity.

The most crucial distinction between the human-induced climate change we are experiencing now and naturally regulated climate change of the past is the persistence and intensity on which they operate. Earth's systems operate on long timelines, thousands to millions of years, as they are driven by natural processes such as the storage of carbon in forests as they grow over centuries, the overturning of ocean circulation, and the multi-million year long process weathering of land as mountains grow and valleys deepen. While slow on human timescales, these processes have regulated the balance of CO_2 (among other things) between the atmospheric, oceanic and continental reservoirs throughout Earth's history. On the other hand, human activity operates on days, years and decades, and has continued unabated through the past 150 years. Our extraction and combustion of oil, coal and natural gas – the fossil fuels, has increased atmospheric CO_2 by 50% since preindustrial times. For comparison this is at a rate roughly 100 times faster than the warming of the last deglacial transition and exceeding any known previous concentrations in the last 14 million years of Earth's history (National Oceanic and Atmospheric Administration, 2022).

Natural processes are buffering the global warming that we are accelerating – the oceans and land have absorbed roughly half of our emissions to date, slowing down the increasing temperatures we are experiencing. But they cannot "fix" the climate crisis for us. As well as increasing the demand on these reservoirs to remove CO_2, we are also reducing their capacity through deforestation, land use change and urbanization. Carbon dioxide that we have already emitted will remain in the atmosphere for up to 1,000 years if left to naturally break down and be absorbed, which creates a significant potential lag time between the slow down or halt of our greenhouse gas (GHG) emissions and a significant reversal of global warming.

Crucially, this means that the impacts of climate change we are already living with are set into Earth's systems for decades to come. Global warming has disrupted natural climate variability, which in turn has introduced volatility into the weather that we experience. Resulting extreme weather events are more frequent, intense and widespread with no signs of slowing down. This is not a case of reaching net zero emissions and resetting the environmental clock. Reducing GHG emissions is the essential and only first step into stabilizing the Earth's natural systems to allow climate recovery in the coming decades and centuries. Let's be clear – we will no longer live in a stable climate, but we can recreate it for future generations and adapt to the instability along the way by taking action today.

To mitigate and adapt for climate change, we first have to understand the natural processes that drive it. Given that our current pace of emissions has pushed planetary boundaries outside of anything we have seen in megaannum, it is important to look for climatically analogous time periods in Earth's history to project our climate future.

Our planet's habitability depends entirely on the fine balance between all of Earth's major systems. Those major systems include the ice caps, land and vegetation, the atmosphere and oceans. Each of these components have varying roles in, and response times to, changing environmental conditions. However, all of these spheres are connected from a global to a molecular level, and it is through this endless network that our environmental conditions ebb and flow. You cannot change one part of the Earth's system in isolation, and so it follows that climate change itself cannot be successfully tackled with a disconnected and siloed approach. Any organizational strategy targeted at building resilience to climate risk must have this unified theory at its foundational core.

Extensive ice sheets at the north and south poles are powerful reflective surfaces, which prevent large amounts of the incoming solar radiation being absorbed into the oceans and land that they cover. The release of cold water from polar regions is critical in driving ocean circulation (and therefore regulating regional weather), as well as trapping and storing CO_2 away from the ocean's surface. Ice sheets also store huge volumes of water, which are estimated to be equivalent to more than 60 m of global sea level rise if melted. Nearly 70% of the world's population live within 100 km of a coastline and even 2 m of sea level rise would displace an estimated 1 billion people by 2050; ice sheet instability is a serious concern in climate change. The response time of the ice sheets – or cryosphere – is slow compared to the atmosphere and oceans and can generally be considered a longer term driver in climate. It would take hundreds or thousands of years to fully melt the polar ice caps, but equally halting the melting process is not instantaneous. Even if we significantly curb emissions in the coming decades, more than a third of the world's remaining glaciers will melt before the year 2100 with an estimated 0.5–1.0 m sea level rise now effectively guaranteed.

Continents are complex systems interacting with climate on several timescales through reservoirs such as rock, soils and vegetation. Carbon dioxide is stored and released naturally on long timescales (millions of years) through the breakdown of rock and the build up and burial of sediments; a process called weathering. Soils are extremely high in organic matter that is released as CO_2 and CH_4 when they are disturbed, generally through land use changes such as peatland draining, agriculture and deforestation, also leading to top soil erosion. Organic matter in frozen soils, known as permafrost, is a huge carbon reservoir, estimated globally at 1,500 gigatons. This is roughly twice as much carbon as the atmosphere contains currently, or the equivalent of 40 times our global CO_2 emissions in 2022. Rising temperatures, which are amplified in the polar regions, are increasingly destabilizing these ancient large carbon stores through ice melt, many of which have been stable for more than 700,000 years.

Natural ecosystems like forests, grasslands, wetlands and mangroves act as carbon sinks, absorbing and storing CO_2 from the atmosphere. Forests, in particular, are vital for carbon sequestration. When trees photosynthesize, they absorb CO_2 and release O_2, storing the carbon in their biomass and in the soil. They also regulate local climate, and their strategic restoration can help drive positive changes in weather patterns, such as providing a moisture corridor to convect rain inland from coastal regions to areas suffering from drought.

The atmosphere and its GHGs including CO_2, CH_4, nitrous oxide, fluorinated gasses and water vapor insulate Earth, maintaining a warm enough environment to support life. However, it is the massive accumulation of additional GHGs compared to pre-industrial levels, specifically CO_2 and CH_4, that are trapping increasingly more heat in and warming the atmosphere. Currently, atmospheric circulation distributes heat from the equator poleward in discrete cells that influence regional climate and weather patterns. It is upon the resulting variety of environments such as forests, grasslands and deserts, that global food supply has been built to depend; from coffee production in the "Bean Belt" of the equatorial tropics to the vast grain supply of the mid to high latitudes. Regional dependence on specific climatic conditions underscores the vulnerability of global food supply to shifts in climate patterns. The same can be said for many other examples from travel and tourism, textiles and apparel, natural resources and raw materials and the building and construction industry. Disruption to the composition and circulation of the atmosphere through warming air temperatures is occurring on a human timescale with immediate consequences of more extreme environmental conditions. It is also triggering change that could last for centuries such as melting ice sheets and transferring heat to oceans, both of which also contribute to sea-level rise.

Our oceans have an immense capacity to store and release heat, acting as a natural thermostat. Heat is redistributed around the globe in the global ocean conveyor belt, which in turn regulates regional climates and has a profound impact on weather patterns. An example of this is the stark contrast between the warm, wet UK climate and the subarctic, cool maritime climate of Canada. Both sit at the same latitude but are adjacent to different water masses, 'which form part of the Atlantic Meridional Overturning Circulation (AMOC) system' the warm Gulf Stream follows the west coast of the UK, while the east coast of Canada is bathed by the cold Arctic Labrador current. If the Gulf Stream were to slow down, or even halt due to changing ocean temperatures, the UK may well experience significant cooling. With potential collapse of the AMOC system predicted by the end of the century, the knock on effects would disrupt agriculture, food production, transport, energy demand, and much more.

The interaction between oceans and the atmosphere is one of the most tangible systems connections on short timescales. Direct heat and gas exchange at the ocean surface is extremely effective, which can stabilise the climate to a point through negative feedback loops as we have discussed. However, the same is true for the reverse where heat and CO_2 can build up in a matter of decades driving changes that we are experiencing in real time. Consistent, long-term warming of oceans and air temperatures introduces increasing levels of energy into the climate system, which in turn create more chaotic interactions. The result is more extreme weather; storms, droughts, heatwaves and wildfires that are becoming more intense, lasting longer and getting less predictable.

Associated disruption and damage ripples through the environment, reducing natural capital stocks such as forests, freshwater and soils. Biodiversity loss increases in parallel with environmental degradation, negatively impacting vital ecosystem services such as pollination, water and air purification and carbon sequestration that society depends heavily on.

Just as climate change alters habitats and ecosystems, loss of biodiversity contributes to climate change and intensifies its effects. If current trends continue, an estimated one million animal and plant species will be on the verge of extinction – higher than any other time in human history. Air pollution now contributes to seven million deaths every year and human disturbance of ecosystems is helping infectious diseases spread more easily. The rapid loss of protective coastal habitats through deforestation has put 100–300 million people at an increased risk of floods and hurricanes, which concurrently, will cause more damage as they intensify due to global warming.

The networked nature of Earth's systems result in feedback loops that can either self-regulate change, known as negative feedback, or trigger a runaway process known as positive feedback. Positive feedback activity in the climate system has the ability to exponentially amplify changing climate conditions. In the case of current human-induced global warming, examples of positive feedback are shown in Figure 2.1.

Positive feedback activity in climate change is an immensely powerful process that has the potential to nullify human-led mitigation efforts. This lack of mitigation efficacy is a result of pushing the climate to points in the natural systems from which you cannot easily return once passed. Known as "tipping points," these are quasi-irreversible changes that cannot be halted once they reach a certain stage – such as changes in large scale ocean circulation, or collapse of the Antarctic ice sheets. Both would lead to an unrecognizable planet operating in a completely different way. Events such as these aren't predicted to be irreversibly triggered without extreme global warming of 4°C or more; however, several significant tipping points including Arctic ice sheet collapse and coral reef eradication are already within reach of our current climate change of 1.2°C, with many more tipping points initiating just another 0.5°C away (Armstrong McKay et al. 2022). These critical temperature thresholds of 1.5 and 4°C respectively have become crucial reference points in modern policymaking, which we will explore in detail, later in this chapter.

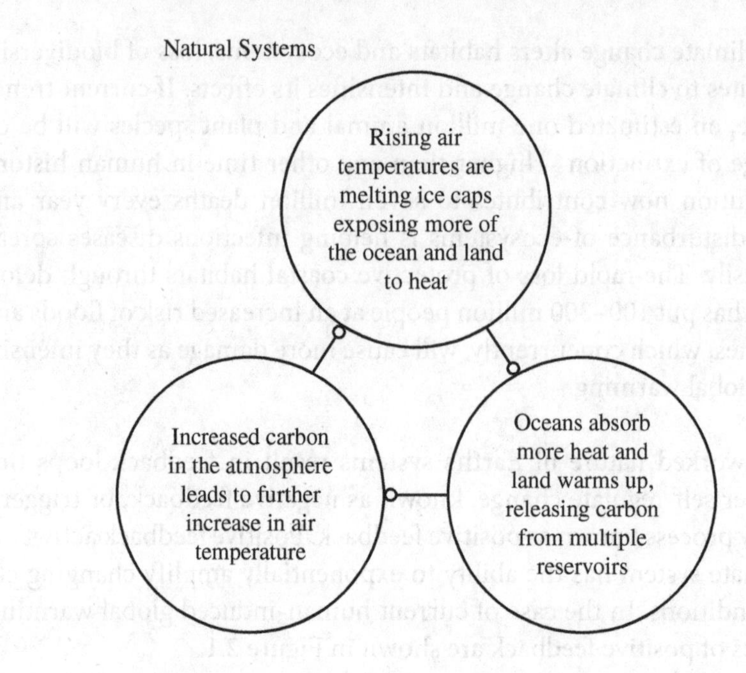

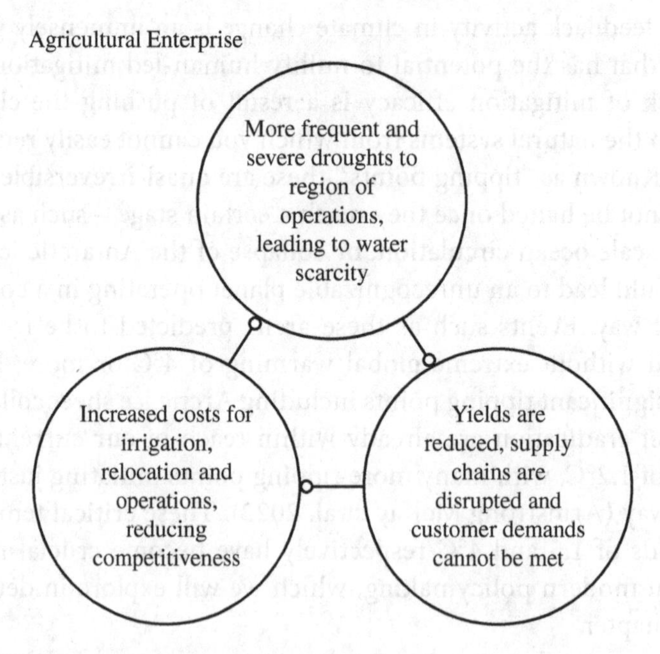

Figure 2.1 Examples of positive feedback loops.

On natural timescales, Earth's climate has been influenced by both positive and negative feedback systems. These mechanisms have shifted the planet between different boundary states and regulated fluctuations within these states. Historically, positive feedback loops, triggered by events such as prolonged volcanic activity, global-scale tectonic reorganisation and even asteroid impacts, have led to shifts between "greenhouse" conditions of extreme global heat and "icehouse" states where much of the planet was frozen. Within these macro-states, negative feedbacks have naturally regulated more predictable drivers, like changing solar radiation levels as Earth's orbit cycles through its elliptical patterns, but they have not significantly altered the status quo. For example, over the past 2.4 million years, Ice Ages have cycled rhythmically through glacial and interglacial phases alongside changing solar radiation levels and steadily rising and falling CO_2 levels. These natural climate changes have occurred within the timeframe and capabilities of the land, oceans, ice sheet and atmosphere's responsiveness, allowing for rebalancing along the way.

Climate tipping points have not yet been witnessed on human timescales, but carry immeasurable implications for our survival. An understanding of how our climate may operate through such large changes, can only come from historical records that pre-date humans. The study of historical climate change - known as paleoclimatology - reconstructs past environmental conditions by using records ranging from satellite observations and historical accounts, to geological archives such as ice cores, tree rings and marine fossils. Some of the natural records are based on direct evidence such as ice cores that preserve atmospheric gas concentrations in tiny bubbles buried deep in glaciers. The vast majority of natural records dating back further than a million years are used as proxies to create indirect reconstructions, i.e. varying analyses of certain elements of these archives can infer environmental conditions at the time where direct evidence is absent. Examples of this include geological records that can track sea level rise and fall, changing fossil pollen records that can show how the vegetation of an area has changed with time (and thus climate) and changing chemical signatures of deep ocean sediment cores that can trace changes in ice sheet growth and decline and alterations in deep ocean circulation patterns.

Alongside current weather and climate records, long-term reconstructions of evolving climate conditions provide essential reference data for modeling future climate change. There are different types of climate models, but the most important ones for understanding global climate trends are called General Circulation Models, or GCMs. These models allow scientists to study how changes in one part of the climate system affect the entire planet. To ensure accuracy, models are fed with climate observations from satellites, weather stations and other monitoring systems. This real-time data helps the models adjust and fine-tune their projections based on what's happening in the world right now.

In addition to the complex scientific data from past and current climate, human activity can also be accounted for within GCMs. Most familiar of these anthropogenic inputs are the emissions scenarios, regularly used in climate policy such as the Paris Agreement. Emission scenarios are essentially "what-if" scenarios that help us understand how different levels of GHG emissions might shape our future climate. They represent hypothetical future trajectories of GHG emissions based on projections of population growth, technological development, economic trends, energy use patterns and policy decisions.

Emissions scenarios have undergone significant evolution over time to provide a more comprehensive and nuanced understanding of future climate trajectories. Starting with basic assumptions about constant GHG concentrations in the early days of climate modeling, the Intergovernmental Panel on Climate Change (IPCC) introduced the IS92 scenarios in 1992. These early scenarios considered a range of emission trajectories based on economic growth, population and technology assumptions. In 2000, the IPCC released the Special Report on Emissions Scenarios (SRES), which introduced a more diverse set of scenarios based on different storylines depicting future societal and technological developments. The introduction of the Representative Concentration Pathways (RCPs) in 2014 provided a shift toward focusing on concentrations of GHGs, offering a clearer link between emissions and climate impacts. Concurrently, the Shared Socioeconomic Pathways (SSPs) were introduced to incorporate a wider range of societal and economic factors.

Emissions scenarios span a range of plausible future pathways that depict different levels of GHG emissions. Low emissions scenarios target minimal global warming of ~1.5°C, assuming aggressive efforts to reduce emissions through widespread adoption of clean energy technologies, energy efficiency improvements and significant policy changes. At the other end of the spectrum are "Business as Usual" (BAU) scenarios, which assume that current trends and practices continue without meaningful changes. The limited global efforts to curb GHG levels results in the mean global temperature increasing by more than 4°C, with extremely dangerous implications for ecosystems, social and economic security world-wide.

In addition to GCMs, an important set of models are Integrated Assessment Models (IAMs). These are comprehensive tools used to evaluate the interactions between human and natural systems in the context of climate change. These models integrate data from multiple disciplines, including economics, energy systems, land use and climate science, to provide a holistic view of potential future scenarios. The primary objective of IAMs is to assess the environmental, economic and social implications of different climate policies and transition pathways. By simulating various policy interventions and technological developments, IAMs help decision makers understand the trade-offs and synergies involved in transitioning to a low-carbon economy. IAMs typically encompass several core components: energy production and consumption, GHG emissions, climate change impacts, economic growth and land-use changes. These components are interlinked within the model to reflect the complex feedback loops and dependencies between human activities and the climate system. IAMs are crucial for identifying cost-effective strategies for reducing emissions, setting targets for renewable energy adoption and evaluating the long-term sustainability of different policy measures. Their outputs guide decision makers in crafting policies and developing strategies that balance economic development with environmental stewardship, ultimately aiming to mitigate climate change and facilitate a smooth transition to a sustainable future. The National Greening of the Financial System (NGFS), as an example, has partnered with leading climate scientists to develop IAMs that model different transition scenarios. These range from "orderly" scenarios assuming necessary action is taken early and "disorderly" scenarios with delayed and divergent actions across different countries to "hot house world" and "too little too late" scenarios with insufficient global action and elevated risks.

Emissions scenarios and model projections are a vital tool for policymakers because they provide valuable insights into the potential consequences of different policy choices as well as a common framework for understanding and comparing the impact of different policy measures across countries. Notably, they play a crucial role in tracking progress toward the overarching goals of the Paris Agreement, a landmark international treaty aimed at combating climate change, which we explore further in Chapter 5. Signed in 2015, the agreement sets a framework for countries to collectively work toward keeping global temperature increases well below 2°C above pre-industrial levels, and ideally limiting the increase to 1.5°C in recognition of the scientific evidence previously discussed. The low-emissions scenarios that follow this trajectory are often referred to as "Paris aligned."

As a baseline, organizations will need to show how they are part of the solution. They will need to define robust targets and develop credible adaptation plans for the transition that set out how they can meet their commitments and play a meaningful part in an essential economy-wide transformation. It forms a significant, inevitable and essential challenge, while also presenting commercial opportunity. The most effective, financially-efficient decisions will be those that (i) are grounded in the scientific information that forms the cornerstone of growing policy and regulatory activity, (ii) consider a range of possible scenarios and (iii) take a proactive approach to addressing climate risk through meaningful action. The challenge lies in deciphering complex international regulatory frameworks, understanding the implications and projections of increasingly volatile climate events to your organization and translating all of this siloed information into decision-useful information. The opportunities for those who can embed strategic climate resilience into their operations range from ensuring supply chain resilience, lowering costs and enhancing brand reputation to unique competitiveness, seizing market opportunities in sustainability.

References

Armstrong McKay, D.I. et al. (2022). Exceeding 1.5°C global warming could trigger multiple climate tipping points. *Science* 377 (6611): https://www.science.org/doi/10.1126/science.abn7950.

National Oceanic and Atmospheric Administration. (2022, 3 June). Carbon dioxide now more than 50% higher than pre-industrial levels. https://www.noaa.gov/news-release/carbon-dioxide-now-more-than-50-higher-than-pre-industrial-levels

CHAPTER 3
CLIMATE RISK AND OPPORTUNITY

Summary

This chapter underscores the urgency to address climate risks and opportunities comprehensively. It highlights the need for global cooperation, the significance of standardized reporting, and the transformative potential of climate intelligence in shaping strategic responses to climate ambitions.

We explore the critical realm of climate risk and opportunity within the context of the rapidly changing business landscape. The multi-faceted concept of climate risk includes physical, nature and transition risks, each with its own set of challenges and consequences and all of which can have far-reaching impacts on economies, environments and societies. The interconnectedness that exists across the spectrum of climate risk underpins our need for comprehensive strategies that address all domains including social, economic and environmental.

In the untapped potential for climate opportunity, we explore the ability for businesses to make positive contributions to society, finance and the environment. Investments in renewable energy, sustainable infrastructure and circularity not only pave the way for a greener future but also yield economic benefits, including cost savings and access to new markets.

In recognizing the global significance of climate risk management, international agreements such as the Paris Agreement and initiatives like the Task Force on Climate-related Financial Disclosures (TCFD) now fully incorporated into new International Financial Reporting Standards (IFRS), sustainability disclosures are supporting standardized climate-related reporting, which enhances transparency and accountability in climate action. They are also a foundational step for informed decision making for businesses, who in turn play a pivotal role in driving climate action. Through leading by example, influencing policy and fostering innovation, organizations can shape the future of a sustainable climate economy.

In today's rapidly changing climate landscape, the considerations of risk and opportunity have evolved from optional to imperative. Comprehending and addressing climate-related challenges plays a pivotal role in ensuring long-term sustainability and viability of countries, cities, businesses and the environment alike. Equally, opportunities are emerging from the adoption of sustainable practices and navigating a low-carbon economy.

Put in its most simple form, climate risk can be defined as the potential adverse impacts or harm caused to our lives, economies and environment by climate-related events and changes. For organizations, it can be defined in practical terms as the measure of vulnerability to climate-related impacts that have financial consequences, or that may affect various aspects of financial performance. Those consequences could be anything from minor inconvenience to a complete loss of an asset's value or operability. Climate risk encompasses physical, nature and transition risk, all of which include the potential for damages, disruptions and financial losses arising from the direct impacts of climate-related events such as extreme weather and natural resource scarcity, indirect impacts like the cascading disruption to global supply chains, and the broader consequences of the transition to a low-carbon economy.

Physical risks from climate change can be categorized as either event-driven (acute) or longer-term shifts (chronic) in climate patterns (Task Force on Climate-related Financial Disclosures, 2017). Acute risks involve event-driven occurrences like cyclones, hurricanes or floods. Chronic risks, on the other hand, result from sustained shifts in climate patterns, such as prolonged high temperatures leading to issues like sea level rise or heat waves. Chronic and acute risks do not occur exclusively and the combination of these can result in even more severe, compounded risk such as increasingly frequent flooding in coastal regions underpinned by rising sea levels. Transition climate risks are related to the shift away from fossil fuels and other greenhouse gas (GHG)-emitting activities, largely in response to changing regulatory pressures. The cost of decarbonizing assets and operations is a transition risk, yet businesses that fail to decarbonize also face other types of transition risk such as reputational loss and loss of market share. We look at these different facets of risk in more detail in Chapter 4.

Climate risk intersects with nature as it involves the potential harm and adverse impacts stemming from climate-related events and human-induced damage and changes that directly affect ecosystems and biodiversity. Biodiversity loss, habitat degradation, deforestation and other ecological factors are all accelerated by or a result of extreme weather events and human activity. Addressing climate risk is crucial for mitigating the negative consequences on nature, and, in turn, managing nature risk is essential for building resilience to climate change impacts. The interconnectedness of these risks highlights the importance of adopting comprehensive management strategies that address both climate and nature-related challenges.

Climate risk is multifaceted and as such, the management strategies that organizations, communities, or countries employ cannot be based on a single aspect of the challenge. Take the example of supply chain disruption. On the one hand, acute physical risks, such as extreme weather events disrupting transportation and logistics, pose immediate threats to supply networks. These events can result in direct damage to assets, delays in production and interruptions in the flow of goods. The impact of biodiversity degradation will create challenges in natural resource availability, disrupting supply of raw materials. On the other hand, businesses are confronted with transition risks linked to the changing sustainability paradigm. Technology risks arise from advancements or changes that can impact the efficiency and resilience of existing supply chain processes. Policy and legal risks come to the forefront as regulations tighten around environmental standards, shaping how businesses source materials and conduct operations. Reputational risk is also emerging as a critical consideration, with consumers and investors increasingly valuing environmentally conscious practices.

While climate change poses significant challenges, there are opportunities for positive social, financial and environmental impact. Investments in renewable energy, sustainable infrastructure and green technologies contribute to a cleaner, more resilient future. Job creation in renewable energy sectors enhances employment opportunities, fostering economic growth. Initiatives addressing climate adaptation and mitigation bolster community resilience and social well-being. Biodiversity conservation and sustainable land use practices present avenues for environmental stewardship. The transition to a low-carbon economy encourages innovation, creating markets for new products and services.

Circularity stands out as a pivotal pillar for climate opportunity because it tackles emissions beyond just energy usage. Its core principles – eliminating waste, circulating materials and regenerating nature – directly address a significant chunk of emissions arising from resource extraction, production and waste management. By extending the lifespan of materials and minimizing extraction, circularity slashes emissions across the entire value chain. This approach goes beyond mitigation, allowing us to retain the embodied energy in materials, reduce pollution and even sequester carbon in nature. With its potential to cut emissions drastically and promote sustainability, circularity plays an important role in building a greener, more resilient future.

For organizations in particular, there are strategic advantages in a transitioning global economy. These opportunities include resource efficiency and cost savings achieved through improvements in sourcing, production and distribution processes. Innovation and technology play a crucial role in this transition, encompassing solutions like efficient heating, circular economy models, renewable energy, water usage, treatment solutions, waste reduction and electrifying transportation. Shifting toward low-emission energy sources, such as wind, solar and others, is emphasized for meeting emission reduction goals, offering potential savings in annual energy costs and future carbon taxes. It offers an avenue remaining more independent of geopolitical turbulence, which provides a level of energy price stability we have not had access to historically. Changing consumer preferences offer opportunities to develop new low-emission products and services. Proactively exploring new markets and types of assets enables diversification, while promoting climate resilience involves adapting to change, managing risks and seizing opportunities related to efficiency and product development.

Embracing climate opportunities not only benefits an organization's top and bottom line, but also aligns with investors' growing focus on sustainability. Companies leading in climate-conscious strategies not only contribute to a sustainable future but also present attractive prospects for financial growth, with strategies aimed at enhanced long-term value, operational efficiency and reduced costs as well as reduced policy risk and increased productivity (Henisz, Koller and Nuttal, 2019). As climate considerations gain prominence in investments, climate-aware companies are well-positioned to attract capital and secure robust financial support. This alignment fosters a climate-centric approach, appealing to conscientious investors seeking positive impact and strong returns.

International recognition for the role of institutions, legislation, and procedure in collective climate risk management has grown significantly following the establishment of the United Nations Framework Convention on Climate Change (UNFCCC) in 1992. To date, climate change has now been incorporated into more than 2,800 laws and policies across the world's countries (Climate change laws of the world, n.d.). However, despite this progress, total emissions have continued to rise every single year. We haven't yet managed to reduce our CO_2 concentrations by a single point. By promoting accountability, transparency and global cooperation, climate framework legislation such as the UN Sustainable Development Goals and the Sendai Framework for Disaster Risk Reduction provide a systematic and coordinated approach for national and international efforts to mitigate GHG emissions, adapt to the impacts of climate change, and promote sustainable practices. We explore the international climate policy landscape in further detail within Chapter 5.

While an essentially complete global consensus exists on the need for emissions reduction, mitigation is fraught with complexity. A significant percentage of GHG accumulation can be attributed to the historic industrialization and economic development of a subset of countries. The modern economic and social success of developed countries is often underpinned by a long-term, historic dependence on fossil fuels, which has been a dominant driver in global warming. This has created a sense of inequity, as developing nations, with lower historical emissions, face the brunt of the subsequent climate impacts despite contributing less to the problem. Many emerging economies now also face the complex task of balancing the need for economic growth with the imperative to reduce emissions.

In addition, the implementation of effective carbon mitigation strategies is confronted by a multitude of challenges. These include technological barriers, such as the high costs and uncertainties associated with new technologies, and the need for consistent innovation. Investment obstacles, particularly for developing nations and smaller enterprises, stem from the substantial capital required for renewable energy and emissions reduction initiatives. Policy and regulatory uncertainty pose further challenges, as the absence of clear incentives and frequent policy changes can discourage long-term private sector engagement. Infrastructure limitations must also be overcome, as transitioning to a low-carbon future necessitates significant changes in energy infrastructure, such as the development of renewable energy generation and upgraded power grids.

Mitigating emissions is crucial, but adapting to the impacts of climate change is also a priority in climate risk management. Even if increasingly ambitious (and unlikely) action is taken to reduce our GHG concentrations to preindustrial levels, residual climate risks and volatility are locked into the environmental system for decades to come. In essence, things will get worse before they get better – even if we were to reach net zero tomorrow. There is no silver bullet solution to the climate change we have driven to date, which is why concurrently building resilience into our economies and societies for the long term is so critical. Again, a global imbalance exists with developing nations facing challenges in implementing measures due to resource constraints, particularly financial, and weak infrastructure while also being at an inherently increased level of vulnerability to climate risk. As damages and economic losses from climate change continue to spiral and costs and risks increase, failing or delaying implementation of adaptation causes more severe impacts on trade flows and geo-concentrated supply networks, and pushes social and environmental resilience toward its limits. In some cases, hard limits can be reached where the capacity to adapt is diminished past a point of no return, such as permanent resource loss, ecological extinction or sea-level rise. As time progresses, these hard limits become closer and more numerous with delays in climate action resulting in fewer options to recover them or adapt to changes.

Climate finance is essential in the mobilization and allocation of funds to support projects and initiatives that contribute to mitigating GHG emissions, adapting to the impacts of climate change and promoting sustainable development. However, the current state of climate finance presents several challenges. The drop in multi- and bilateral finance for adaptation since 2020 has raised concerns about the availability of funds to address the pressing needs of climate adaptation efforts. This decline in funding exacerbates the existing financing gap, which stands at a staggering 10 times the available funds annually. This annual financing gap, currently estimated to be in the range of $215 billion to $388 billion, poses a significant obstacle to effectively responding to climate-related risks and challenges and will continue to grow each year.

To compound the issue, while efforts have been made to address climate finance, they often fall short of the necessary scale. For instance, the Loss and Damage fund agreed upon at COP28, with its $700 million pledge, covered less than 0.2% of the funds required at the time to adequately address loss and damage resulting from climate change impacts.

This underscores the urgent need for more substantial commitments and innovative financing mechanisms to bridge the financial divide and ensure that vulnerable communities and nations receive the support they require.

Climate finance can realize the diverse opportunities presented by the transition to a low-carbon economy, investing in renewable energy, enhancing resilience to climate-related risks and assisting developing nations in coping with the adverse effects of climate change. Further opportunity presents itself in the creation of a level playing field to share the investment costs across all industries and countries through cooperation. However, delayed or differentiated efforts will likely lead to "go-it-alone" approaches that prove fruitless. Climate change is a transboundary challenge, and can only be successfully managed with equally global coordination. As such, a collective and cooperative approach to climate finance is essential to ensure effective mitigation and adaptation measures that benefit both developed and developing nations – directly and indirectly.

The heightened awareness of the profound implications of climate change has triggered a paradigm shift in how businesses and investments are evaluated. At the heart of the market forces driving increasing pressure on organizations – especially large publicly listed companies – is the issue of financial materiality of climate change. The inflation adjusted financial costs of climate and weather-related extreme events in the United States has gone from $21 billion per year in the 1980s to more than $120 billion per year over the past five years (National Oceanic and Atmospheric Administration, 2024). As the consequences of climate change become more tangible and widely acknowledged, stakeholders such as investors, customers and regulatory bodies are recognizing the need for transparency, especially on how climate change is going to impact – both positively and negatively – the financial performance of enterprises. Investors, aiming to make informed decisions, are seeking comprehensive insights into the climate-related risks and opportunities facing businesses and how they plan to address these. Customers, too, are increasingly mindful of the environmental impact of the companies they support, influencing purchasing and investment decisions. Concurrently, regulatory bodies worldwide have been taking proactive measures to align financial reporting and trading with the imperatives of sustainability and create standardized frameworks for disclosing climate-related financial risks and opportunities.

By introducing standardized guidelines, regulatory bodies such as the Financial Stability Board (FSB) are aiming to enhance the quality and relevance of climate-related information disclosed by businesses through frameworks such as the Task Force on Climate-related Financial Disclosures (TCFD) and the Task Force on Nature-related Financial Disclosures (TNFD) initiatives. The International Financial Reporting Standards Foundation (IFRS) formed the International Sustainability Standards Board (ISSB) that recently released two climate and sustainability standards as they look to harmonize reporting requirements. In the European Union, more than 49,000 organizations (covering more than 75% of European business revenue) will be subject to the recent Corporate Sustainability Reporting Directive (CSRD), which will require companies to disclose double materiality – that is both the sustainability and climate-related factors that affect company performance as well as the company's impact on the environment. As these regulatory frameworks gain traction, they not only promote accountability but also empower stakeholders to make more informed decisions, encouraging a collective shift toward sustainable and climate-resilient business models.

In the face of growing regulatory, financial and consumer pressures, organizations are increasingly recognized as pivotal actors in the global response to climate change. There is an opportunity for businesses to act flexibly, adapt quickly, adjust their operations, integrate carbon pricing into products and refine services to align with sustainability goals without the need for lengthy legislative processes. In the interests of remaining financially secure within a competitive landscape, businesses are a significant driving force for change. Organizations that lead by example and signal their commitment to climate action will competitively lead the pack in an emerging climate economy and can inspire others within their industry or sector to follow suit. This can trigger a domino effect, with more companies adopting climate-friendly practices and driving industry-wide change. Collectively, they can advocate for supportive climate policies and regulations, engaging with governments and policymakers to shape climate-related legislation and a regulatory environment that encourages sustainable practices and innovation. There is also the significant opportunity for organizations to accelerate progress toward a low-carbon economy by attracting partners, research institutions and innovators interested in working together to develop and implement new technologies and solutions.

The initial wave of climate disclosures has offered organizations an opportunity to baseline their current climate risk, measure GHG emissions, and evaluate their exposure to acute and chronic events, policy risks and more. Akin to a health check, this process is crucial as it provides a foundation for informed decision making, enabling organizations to invest strategically and implement the necessary interventions to address climate-related challenges effectively.

While the initial disclosure efforts provide a high-level assessment, the analytical unit for risk assessment or policy evaluation typically operates at a high spatial resolution. This means that the assessment is conducted on a broader scale, possibly at a regional or national level, and may not capture risks and opportunities at the operational or property-specific level. This limitation can result in the oversight of climate-related risks that exist within networks of assets or financial instruments tied to these assets. As such, it may not provide a granular understanding of risks and opportunities associated with individual assets or financial instruments.

To address this gap, organizations and policymakers are looking to develop more detailed, business-unit and asset-level assessments and insights to ensure comprehensive management of climate related financial risks and opportunities. This shift toward aggregated asset-scale analysis is helping organizations identify and act upon both risks and opportunities across their portfolios, leading to better performance, effective climate resilience and sustainable value creation.

There is no single point of responsibility for climate risk. Proactive and collaborative efforts to build resilience and prepare for adaptation are needed at and between every level. Global pressure for climate action has created an urgent and permanent demand for the technology, capital, skills and insights to meet regulatory standards and future proof social, environmental, and economic security. This demand has evolved beyond carbon assessments and now encompasses a broader spectrum of challenges, including physical risk, policy compliance, transition readiness, technological innovation, resource use and nature-based risks. We are beginning to recognise the limitations of siloed and isolated decision making processes. However, understanding climate risk and opportunity also requires multidisciplinary expert knowledge, often including skill sets that are not found in the majority of organizations. New technologies must not only simplify the complex relationships between climate science and business intelligence but also provide decision-useful insights that can drive the financial performance and climate action that needs to be taken.

The push for climate-aligned decision making across all sectors has led to the genesis of a powerful new technology; climate intelligence (CI). Climate intelligence fuses expertise in climate science, artificial intelligence, financial analysis, environmental policy and sustainability practices empowering decision makers with informed choices that align with both environmental goals and long-term profitability. It is fast becoming an essential capability for organizations to navigate the climate crisis, supporting effective decision making across all activities, while remaining rooted in the science that underpins climate action at the highest level. In the rest of Part I, we will deep dive into the different risk types and explore the rapidly evolving CI landscape that is shaping strategic organizational responses to climate ambitions.

References

Climate change laws of the world (n.d.). https://climate-laws.org/

Henisz, W., Koller, T. and Nuttal, R. (2019, 14 November). Five ways that ESG creates value. *McKinsey Quarterly*. https://www.mckinsey.com/capabilities/strategy-and-corporate-finance/our-insights/five-ways-that-esg-creates-value

National Oceanic and Atmospheric Administration. (2024, 8 February). United States summary. https://www.ncei.noaa.gov/access/billions/state-summary/US

Task Force on Climate-related Financial Disclosures. (2017, June). Recommendations of the Task Force on Climate-related Financial Disclosures. https://assets.bbhub.io/company/sites/60/2020/10/FINAL-2017-TCFD-Report-11052018.pdf

CHAPTER 4
A DEEP DIVE INTO RISK TYPES

Overview

In this chapter, we delve into an in-depth exploration of three macro types of climate risks: physical risk (PR), transition risk (TR) and nature-related risk (NR). Each type of risk presents distinct challenges and opportunities that significantly influence strategic decisions within organizations. While some risks are widely recognized, others remain less apparent, yet no less impactful. We begin by understanding these risks individually, revealing their specific impacts and policy landscapes. This chapter is intended to serve as a foundational reference and guide, elucidating the complexities of climate risk and opportunity. We conclude by exploring the vital importance of addressing these risks in an integrated manner to effectively manage and mitigate the broader implications of climate change.

CHAPTER 4A
PHYSICAL RISK

Summary

Physical Risk (PR) encompasses the potential adverse outcomes that businesses and communities face due to climate-related hazards, such as extreme weather events and long-term environmental changes. These risks pose significant threats to infrastructure, supply chains and overall business operations, necessitating a thorough understanding and proactive management approach. This chapter outlines the various types of PR, including acute, chronic and compounding risks, and highlights the drivers behind these risks, from environmental factors to socio-economic influences.

Effective management of PR involves utilizing advanced tools and methodologies, such as scenario analysis, stress testing and geospatial analytics. The chapter also explores the opportunities that arise from these challenges, showcasing innovations in technology, infrastructure resilience and financial instruments. By leveraging these insights and integrating them into their strategies, organizations can enhance their resilience, safeguard their assets and capitalize on the emerging opportunities in a changing climate landscape.

In this chapter we explore physical risk (PR) in detail, looking at the drivers, impacts and management approaches for minimizing the risk and maximizing potential opportunities. Physical risk refers to the potential adverse outcomes that businesses and communities face due to climate-related hazards. These hazards, which include extreme weather events like hurricanes, floods, wildfires and long-term environmental changes such as rising sea levels and prolonged droughts, pose significant threats to the stability and functionality of infrastructures, supply chains and overall business operations. Physical risk will continue to create substantial economic losses in the coming years. With an estimated $38 trillion in global damages projected by mid-century (Kotz, Levermann and Wenz, 2024) the significance of PR cannot be overstated.

Understanding and managing PR is crucial for organizations, as the frequency and intensity of climate-related events continue to rise due to global warming. These costly risks can lead to damaged assets, interrupted production, increased operational costs and heightened insurance premiums. Notably, many low-resilience assets are already integrated into the global economy, which means PR can easily disrupt supply chains, affect workforce availability and diminish market demand. This poses significant challenges to profitability and competitive positioning, underlining the urgency with which adaptation actions must be implemented.

Investors, too, are increasingly aware of the importance of accounting for PR in their portfolios. The potential for asset devaluation and the increased volatility associated with climate-related events can impact investment returns. As a result, there is a growing demand for transparency in how organizations assess and manage their exposure to PR, prompting businesses to integrate climate resilience into their strategic planning and reporting.

Often overlooked are the interdependencies and cascading effects of PR. Physical risks in one sector can trigger ripple effects across other sectors and regions, highlighting how vulnerabilities in one area can exacerbate risks in another. For example, the agriculture sector relies heavily on stable weather patterns. A significant climate event affecting agriculture can lead to food shortages, increased prices and economic stress in dependent industries such as food processing and retail. Similarly, disruptions in energy infrastructure due to extreme weather can have far-reaching consequences for manufacturing, transportation and healthcare services. Such interconnectedness means that PR cannot be viewed in isolation; the potential domino effect of failures across various sectors and geographies must be considered in effective strategic planning.

Importantly, the same is true for the different categories of extreme weather – you must consider a complete picture of all risk signals from heat waves to flooding due to the interconnectedness of Earth's natural systems. Planning for one risk in isolation will not provide adequate protection and will undermine overall resilience. Physical risks can be categorized into acute, chronic and compounding risks, each with distinct characteristics and implications for businesses.

Acute risks refer to sudden, severe weather events that cause immediate damage. Examples include hurricanes, floods and wildfires. These events can devastate assets and operations in a short period, leading to significant financial losses and operational downtime. The costs associated with acute risks are staggering; hurricanes alone have caused an average of more than $300 billion in damages annually in the United States over the past decade. For comparison, the average annual cost in the previous decade was $28 billion (Office for Coastal Management, n.d.). Furthermore, damage from floods globally is estimated to rise to more than $700 billion annually by 2030 (World Resources Institute, 2020) due to socioeconomic growth and climate change. The economic impact of such events underscores the urgent need for businesses to integrate acute risk management into their strategic planning.

Chronic risks are long-term climate-related changes that gradually impact businesses and ecosystems. These include sea-level rise, prolonged droughts and gradual temperature increases. The implications of chronic risks extend beyond immediate damage, affecting business sustainability and operational efficiency over time. Sea-level rise, for example, threatens coastal infrastructure and can lead to the displacement of communities and businesses. Research indicates that a 1m rise in sea levels could result in global economic losses exceeding $14 trillion annually by 2100 (Jevrejeva et al., 2018). These long-term changes necessitate proactive measures to ensure the continued viability of businesses in affected regions.

Compounding risks occur when multiple climate-related events happen simultaneously or in close succession, amplifying their overall impact. These can include scenarios such as a hurricane followed by heavy rainfall, leading to exacerbated flooding, or a heatwave combined with drought conditions, intensifying the strain on water and energy resources. The interplay of compounding risks significantly challenges resilience and recovery efforts, as the cumulative effects can overwhelm existing response mechanisms and infrastructure. In lower-income countries, the impact of climate-related events can be particularly severe. An initial extreme weather event might significantly reduce the resilience of these communities, which often struggle to recover before the next disaster strikes, largely due to a lack of resources. This cycle of vulnerability highlights the urgent need for targeted support and effective resilience-building strategies to prevent compounding impacts that can stall recovery and development. Addressing compounding risks involves a holistic approach to risk management, incorporating comprehensive scenario planning and cross-sectoral coordination to enhance resilience against multiple, interconnected threats.

The primary drivers of PR can be broadly categorized into environmental and socio-economic factors. The important distinction here is that environmental factors drive increasing climate volatility, while socio-economic factors increase vulnerability to PR by determining how exposed and sensitive populations and infrastructure are.

Environmental Factors: Global warming is a significant driver of PR, fundamentally altering the frequency, intensity and distribution of natural disasters. Rising global temperatures cause increased occurrences of extreme weather events such as hurricanes, floods and wildfires by altering the behavior and interactions of our climate system. Warmer ocean temperatures provide more energy for hurricanes, making them more powerful and destructive. Similarly, prolonged periods of higher temperatures can lead to severe droughts and increased wildfire activity.

Socio-economic Factors: Urbanization and economic development significantly influence vulnerability to PR. Rapid urbanization, especially in coastal and flood-prone areas, has increased the exposure of populations and infrastructure to natural disasters. As cities expand and populations grow, more people and assets become susceptible to the impacts of extreme weather events. Economic development can both mitigate and exacerbate PR. On one hand, improved infrastructure and better building codes can reduce vulnerability. On the other hand, increased economic activity often leads to higher emissions and greater environmental degradation, contributing to climate change and amplifying PR.

Physical risk presents a multitude of challenges across economic, environmental and social dimensions. These impacts necessitate a thorough understanding to develop effective strategies for resilience and adaptation.

Extreme weather events can lead to loss of life, physical displacement and economic instability. Communities hit by hurricanes or floods often experience long-term economic downturns, increased poverty rates and disrupted local economies. The broader impacts of PR ripple through societies, affecting everything from housing and infrastructure to public health and education systems. Physical risk is having an increasingly significant impact on the workforce impacting employee safety, health and productivity. Floods and hurricanes can disrupt daily operations, making it difficult for employees to commute to work or even displacing them from their homes. Extreme heat can reduce worker productivity, increase health-related absences and heighten the risk of heat-related illnesses. In the US alone, without any mitigation against the rising heat, climate-related productivity losses could reach $200 billion by 2030 and $500 billion by 2050 (Atlantic Council, 2021). In areas projected to experience higher rises in temperatures such as South Asia and West Africa, the hours of productivity loss are expected to be double those of other parts of the world.

Natural disasters also have profound environmental implications leading to habitat destruction, loss of biodiversity and degradation of ecosystems. Wildfires can devastate forests, destroying wildlife habitats and altering ecosystems. Flooding can lead to soil erosion, water contamination and loss of arable land. These environmental impacts can have long-lasting effects, further exacerbating socio-economic vulnerabilities, continuing a negative feedback cycle of further environmental degradation and complicating recovery efforts.

Alongside the significant challenges, PR presents opportunities for innovation and resilience across various sectors. By leveraging these opportunities, organizations can enhance their resilience, drive sustainable growth and contribute to broader climate adaptation efforts.

Increasing PR has driven substantial technological innovation that can monitor, mitigate and optimize for impacts. Emerging technologies, such as advanced climate modeling, remote sensing and predictive analytics, are becoming increasingly essential. For example, Internet of Things (IoT) sensors and AI-driven analytics can provide real-time data on weather patterns, helping companies anticipate and respond to extreme weather events more effectively. And of course, climate intelligence has become a pivotal tool in business strategy. These innovations not only improve risk management but have opened new opportunities.

Investing in infrastructure resilience is another key area where PR are driving positive change. Durable building infrastructure that can withstand extreme weather conditions is crucial for long-term sustainability. Green building and sustainable design practices offer significant opportunities for enhancing resilience. Incorporating features such as green roofs, rainwater harvesting systems and energy-efficient materials into building designs can reduce vulnerability to physical risks. The Bullitt Center in Seattle is one of the greenest commercial buildings in the world, designed to be resilient to natural disasters while minimizing its environmental footprint. It is almost three times more energy efficient than the highest rated LEED buildings and is designed to last at least 250 years. By adopting such practices, companies can reduce operational risks, lower insurance costs and attract environmentally conscious tenants and investors.

Insurance plays a crucial role in mitigating the financial impacts of PR. Financial instruments, such as catastrophe bonds and resilience bonds, offer ways to manage risk exposure. Catastrophe bonds allow insurers to transfer risks to capital markets, providing additional financial resources in the aftermath of a disaster. These bonds attract investors seeking to diversify their portfolios with non-correlated assets. Resilience bonds, on the other hand, are designed to fund infrastructure projects that reduce vulnerability to PR. By linking bond payouts to resilience-building activities, these instruments incentivize proactive risk management.

Public-Private Partnerships (PPPs) play a pivotal role in enhancing resilience to PR. By combining the resources and expertise of the public and private sectors, PPPs can drive large-scale climate adaptation and risk mitigation projects. For the Rotterdam Climate Initiative in the Netherlands, the city government partnered with private companies to implement innovative water management solutions. The Rebuild by Design program in New York brought together federal and local governments, private firms and community organizations to develop resilient infrastructure in the wake of Hurricane Sandy.

In the agriculture sector, the development of climate-resilient crops and advanced irrigation systems is essential for withstanding extreme weather conditions such as droughts and floods. Leading global organizations such as Bayer are investing heavily in agricultural biotechnology to develop crops that are more tolerant to climate stressors. Precision agriculture technologies are increasingly allowing for real-time data collection and analysis, enabling farmers to optimize water usage and improve crop management practices. Due to rising adoption, the precision farming market is growing rapidly from $7 billion in 2020 to a projected $12.8 billion by 2025.

The energy sector is witnessing substantial investments in renewable energy sources, such as wind and solar power, which not only reduce dependency on fossil fuels but can also enhance certain aspects of grid resilience. The 2022 US Inflation Reduction Act (IRA) represents a landmark policy, providing $369 billion in funding for clean energy and climate initiatives. Clean energy investments are also growing globally, reaching $1.8 trillion in 2023 (BloombergNEF, 2024) and this figure is projected to increase significantly as countries strive to meet their climate targets.

Within the healthcare sector, hospitals and other facilities are increasingly adopting resilient design principles to ensure they can operate effectively during extreme weather events. Following Hurricane Sandy in 2012, the New York University Langone Health facility invested $1.5 billion in flood protection and infrastructure improvements, which included flood barriers and elevated power systems to ensure continuous operation during future storms. Public health initiatives aimed at addressing climate-related health issues, such as heatwaves and vector-borne diseases, are gaining traction, particularly as the World Health Organization (WHO) estimates that climate change will cause an additional 250,000 deaths per year between 2030 and 2050 (World Health Organization, 2023), highlighting the urgent need for adaptive healthcare strategies.

Effective management of PR requires robust assessment tools and methodologies such as scenario analysis, stress testing and geospatial analytics.

Scenario analysis, such as the climate models provided by the IPCC, involves creating detailed simulations of potential PR scenarios to assess their impacts on assets, operations and financial performance. This method helps organizations anticipate various outcomes and develop strategic responses to different risk scenarios. Stress testing is a critical tool for evaluating an organization's financial resilience to PR. It involves simulating extreme but plausible adverse conditions to understand how they could affect the financial health of the business. Industries such as banking and insurance have been at the forefront of employing stress testing. Geospatial data and predictive models play a crucial role in risk assessment by providing detailed, location-specific insights into PR. These tools use satellite imagery, geographic information systems (GIS) and advanced algorithms to predict and visualize the impacts of climate events.

The climate intelligence sector leverages these powerful tools, integrating AI and machine learning into purpose-built platforms to deliver insights on specific physical hazards. Flood risk management, for example, integrates detailed flood modeling, digital elevation models, risk assessments and predictive models to help insurers, governments and businesses understand their exposure. Similarly, wildfire risk intelligence uses real-time monitoring and advanced modeling techniques to predict wildfire behavior and assess potential impacts. These solutions support fire management agencies, utilities and insurance companies in enhancing wildfire preparedness and response strategies. By utilizing these advanced tools and technologies, organizations can significantly improve their resilience to PR, proactively manage potential threats and capitalize on emerging opportunities in a changing climate landscape.

References

Atlantic Council (2021, August). Extreme Heat: The Economic and Social Consequences for the United States. https://www.atlanticcouncil.org/wp-content/uploads/2021/08/Extreme-Heat-Report-2021.pdf

BloombergNEF (2024, 30 January). Global Clean Energy Investment Jumps 17%, Hits $1.8 Trillion in 2023, According to BloombergNEF Report. https://about.bnef.com/blog/global-clean-energy-investment-jumps-17-hits-1-8-trillion-in-2023-according-to-bloombergnef-report/

Jevrejeva, S. et al. (2018). Flood damage costs under the sea level rise with warming of 1.5 °C and 2 °C. *Environmental Research Letters.* https://iopscience.iop.org/article/10.1088/1748-9326/aacc76

Kotz, M., Levermann, A., and Wenz, L. (2024). The economic commitment of climate change. *Nature* 628: 551–557. https://doi.org/10.1038/s41586-024-07219-0.

Office for Coastal Management. (n.d.). Fast Facts – Hurricane Costs. https://coast.noaa.gov/states/fast-facts/hurricane-costs.html

World Health Organization (2023, 12 October). Climate change. https://www.who.int/news-room/fact-sheets/detail/climate-change-and-health#:~:text=Research%20shows%20that%203.6%20billion,diarrhoea%20and%20heat%20stress%20alone.

World Resources Institute (2020, 23 April). New Data Shows Millions of People, Trillions in Property at Risk from Flooding — But Infrastructure Investments Now Can Significantly Lower Flood Risk. https://www.wri.org/news/release-new-data-shows-millions-people-trillions-property-risk-flooding-infrastructure

CHAPTER 4B
TRANSITION RISK

Summary

Transition Risk (TR) refers to the vulnerabilities organizations face as economies shift toward a lower-carbon economy. This risk is driven by significant transformations in legislation, technology, market dynamics and societal expectations. Understanding TR is essential for businesses and investors to navigate the complexities of transitioning to a climate economy and to strategically manage the potential opportunities and impacts on their operations and financial performance.

Navigating the unique complexities of TR requires sophisticated tools and approaches. This chapter explores how various climate data and analytics tools, such as scenario analysis, stress testing and carbon accounting, can inform strategic decision making. We also discuss real-world examples of how different sectors, including real estate, banking, power and utilities, insurance and commodities, are adapting to and capitalizing on the transition to a low-carbon economy. By integrating TR intelligence into their operations, businesses can better manage the uncertainties of climate change and ensure long-term financial stability.

The Task Force on Climate-related Financial Disclosures (TCFD) outlines Transition Risk (TR) as stemming from the process of adapting to a lower-carbon economy, which is influenced by changes in policy, technology, markets and reputation. Each of these risk categories plays a crucial role in shaping how TR manifests across different industries and sectors. Understanding these distinctions is vital for businesses and investors, as it allows for a more nuanced approach to managing risks associated with the economic shift toward sustainable practices. Here is a general definition of each:

Policy Risk: arises from changes in laws, regulations and policies that aim to address mitigation of and adaptation to climate change. This includes legislation that imposes stricter emissions targets, carbon pricing and other regulatory measures intended to curb greenhouse gas emissions. Practical implications may include increased compliance costs, enhanced reporting requirements and potential operational restrictions that could affect profitability.

Market Risk: involves changes in supply and demand for goods and services due to the shift toward a low-carbon economy. This risk is often driven by changing consumer preferences, increased demand for green products or shifts in resource availability due to climate policies. Companies might face risks related to declining demand for fossil fuels, growth in renewables or disruptions in supply chains as markets adjust to new standards and expectations.

Technology Risk: stems from the transition to technologies that reduce emissions and are more aligned with a sustainable economy. This includes the risk that new technologies will disrupt existing business models or that companies will face high costs transitioning to these new technologies. There's also the risk of technological obsolescence, where current technologies or assets become outdated or less competitive compared to newer, more efficient solutions.

Reputation Risk: involves the potential damage to a company's reputation if it fails to meet stakeholders' expectations regarding climate action. This can impact customer loyalty, investor confidence and overall corporate image. As public awareness and concern about climate change increases, companies perceived as not contributing to mitigation efforts or engaging in practices harmful to the environment can experience boycotts, divestments or other negative impacts on their brand value.

TR is uniquely complex because the different risk facets do not exist independently of other macroeconomic or extraneous risks. So while TR often relates to an organization's exposure to specific facets of a decarbonization scenario, the actual impact may be triggered by events that are not directly related to climate change. For instance, the recent energy crisis, exacerbated by geopolitical tensions, highlights the profound implications of TR due to the interconnected nature of global economies and energy systems. Furthermore, understanding the full spectrum of TR requires a detailed examination of different emissions scopes, which are themselves also susceptible to both climate-related and broader economic disruptions:

- *Scope 1* emissions are direct emissions from owned or controlled sources, such as company-owned vehicles and facilities.
- *Scope 2* emissions are indirect emissions from the generation of purchased electricity, steam, heating and cooling consumed by the reporting company.
- *Scope 3* emissions encompass all other indirect emissions (not included in Scope 2) that occur in the value chain of the reporting company, including both upstream and downstream emissions.

These categories are vital for companies to identify where they can reduce emissions and how they might be affected by regulatory changes, market shifts and technological advancements. For instance, an increase in carbon pricing would directly impact Scope 1 and 2 emissions costs, while changes in consumer preferences and new regulations could significantly affect Scope 3 emissions.

In addition to the energy related costs, TR presents several other tangible impacts on organizations and individuals alike. Future carbon taxes are a looming financial threat that could cause asset stranding; as governments worldwide impose stricter measures to curtail carbon emissions, assets reliant on fossil fuels may lose value or become obsolete. This scenario forces companies to either write off these investments prematurely or incur significant costs to adapt to new regulations. The shift to low-carbon alternatives also extends to residential settings, particularly in heating solutions. Transitioning from traditional gas boilers to electric heat pumps involves considerable initial expenses. This not only affects homeowners by increasing their financial burdens but also poses risks to banks and financial institutions.

Importantly, the associated risk of physical infrastructure due to increased electrification cannot be overlooked. As more sectors rely on electricity over other energy forms, the demand for a resilient electrical grid surges, highlighting the physical risks (PRs) tied to electrification, such as the vulnerability of electrical infrastructure to extreme weather events, which can cause widespread disruptions. Companies must invest in enhancing the resilience of their infrastructure, integrating considerations of both physical and TRs into their strategic planning.

Transition risk is fundamentally different from PR in that it doesn't necessarily increase as a result of the lagging effect of rising emissions concentrations. Understanding TR involves comprehending the multiple dynamic factors that drive its impact across different sectors. Policy decisions, such as the introduction of carbon pricing, renewable energy subsidies, and emissions penalties, can drastically shift market dynamics overnight, profoundly affecting certain industries. The manner and speed at which these regulations are implemented – termed regulatory responsiveness – also play a crucial role. Rapid changes can disrupt established business models, while slow enforcement leads to uncertainties that can hinder long-term planning and investment.

Market responses further complicate things. As consumer preferences shift toward more sustainable products and services, companies that are slow to adapt may find themselves at a competitive disadvantage, potentially reshaping entire market standings. Additionally, the availability and advancement of technology are pivotal. Technological innovations that facilitate the shift to a low-carbon economy, such as renewable energy technologies or energy-efficient systems, can drastically alter the business landscape. However, the rate at which these technologies advance and are adopted can vary, adding another layer of risk to companies navigating the transition to a more sustainable operational model.

Addressing the drivers of TR necessitates a strategic approach due to the inherent uncertainty in policy outcomes, policy durability and market responses. Rather than attempting to predict specific policy futures, it is prudent to use a standard set of scenarios as benchmarks. These benchmarks help quantify the range of possible outcomes and impacts, providing a structured framework for risk assessment and strategic planning. For a more tailored analysis, organizations can engage in custom scenario modeling. By working with expert consultants and using bespoke integrated assessment models, businesses could explore specific scenarios in greater detail. This approach is particularly valuable for organizations with dedicated policy and strategic foresight teams that need to prepare for or react to specific policy outcomes, enabling them to strategize and adapt with a clearer understanding of potential future environments.

While there are obvious risks associated with transitioning to a low-carbon economy, the shift also presents numerous opportunities for businesses positioned to capitalize on these changes. Change not only necessitates adaptation but also opens avenues for innovation, cost reduction and the creation of new business models that can deliver substantial economic benefits.

Within the real estate sector, the transition to a low-carbon economy opens multiple avenues for innovation and financial benefits. Real estate companies are increasingly aligning with green finance principles by issuing green bonds, which often attract lower interest rates due to their appeal to environmentally-conscious investors. For example, Lendlease's issuance of 500 million euros in green bonds highlights investor confidence in sustainable projects. The globally recognized waterfront precinct, Barangaroo South in Sydney, developed by Lendlease, not only achieved carbon-neutral status but also significantly increased its valuation, from an initial AUD 6 billion to more than AUD 14 billion, through efficient energy use, water savings and waste reduction strategies. By coordinating with tenants, lenders and utilities, real estate firms can finance energy efficiency (EE) upgrades and electrification projects that not only reduce operational costs but also enhance the asset's value and appeal to a growing demographic of tenants who prioritize environmental sustainability. Meanwhile, Hines's Salesforce Tower in San Francisco, which has the highest Leadership in Energy and Environmental Design (LEED) certification, incorporates advanced water recycling systems and a high-efficiency ventilation system. These features make the building highly desirable, thereby supporting higher rental prices and lower interest rates on green bonds, underscoring the financial benefits of sustainable building practices.

Recognizing the financial stability afforded by energy savings, banks are innovating their product offerings to encourage sustainable housing practices. Energy savings typically result in lower default rates and improved loan performance, prompting Barclays to pioneer the green mortgage initiative. This program offers discounted mortgage rates to those purchasing energy-efficient homes or making green upgrades. Similarly, ING now provides lower interest rates to borrowers investing in energy-efficient properties or committing to specific sustainability improvements, reflecting the bank's broader commitment to environmental sustainability. These initiatives demonstrate how financial products can evolve to encourage eco-friendly practices, aligning financial incentives with positive environmental outcomes. For example, Barclays' green mortgage initiative has shown that properties with higher energy efficiency ratings are less likely to default, leading to more stable loan portfolios. Additionally, ING's approach has seen increased uptake of energy-efficient mortgages, driving both environmental benefits and customer engagement.

The power and utilities sector is proactively embracing the transition to a low-carbon economy through strategically designed EE programs and electrification subsidies tailored to specific households or neighborhoods. Pacific Gas and Electric (PG&E) in California offers a variety of rebates for residential customers who choose to install energy-efficient appliances and solar panels, effectively reducing energy use and supporting grid stability. Similarly, Duke Energy in North Carolina has implemented dynamic pricing models that incentivize customers to use electricity during off-peak hours, enhancing overall energy efficiency and easing grid demand. In the UK, the Green Homes Grant scheme provided homeowners and landlords with vouchers to install energy-efficient improvements, ranging from insulation to heat pumps, aiming to reduce energy usage across the nation. Utilities such as E.ON UK have partnered with local governments to further promote these upgrades, combining private and public efforts to accelerate the transition toward more sustainable energy consumption patterns.

Utilities are also increasingly collaborating with regulators to ensure that infrastructure upgrades not only comply with current standards but are also resilient and adaptable to future climate scenarios. In Germany, E.ON has collaborated with regulatory bodies to implement a wide-scale smart grid initiative. This project integrates a higher percentage of renewable energy sources into the national grid and employs advanced energy storage technologies to enhance grid stability and efficiency. Such partnerships are pivotal as they not only meet the stringent EU regulations aimed at reducing carbon emissions but also pave the way for a more resilient energy infrastructure capable of handling the increasing load from renewable sources.

Insurance companies like Swiss Re are at the forefront of adapting to climate-related risks by refining their underwriting processes to favor less risky customers and sectors, thereby reflecting a deeper understanding of climate-related vulnerabilities. Swiss Re has been particularly proactive in developing parametric insurance products, which offer quick, predetermined payouts triggered by specific climate events. Such measures not only improve risk management but also demonstrate how insurance models are evolving in response to the complexities of climate change.

In the commodities sector, companies like SSAB (Swedish Steel AB) in Sweden are leading the way with innovative practices that significantly enhance physical resiliency and effectively manage transition risks. SSAB has pioneered the production of green steel, which is manufactured using hydrogen instead of coal, drastically reducing the carbon footprint associated with steel production. This green steel not only meets increasing regulatory and consumer demand for sustainable products but also commands a premium price in the market due to its reduced environmental impact. Such initiatives showcase how commodities businesses can adapt their supply chain processes to align with a low-carbon economy while capturing new market opportunities.

For organizations, navigating TR requires sophisticated tools that not only mitigate potential risks but also optimize opportunities. These tools are indispensable for companies aiming to reduce risk and capitalize on the transition to a low-carbon economy by leveraging data analysis and strategic insights. Climate intelligence provides a strategic framework for addressing these risks through various data analysis tools such as scenario analysis, stress testing, and carbon accounting. These insights are indispensable for informing strategic decision making and ensuring that businesses are well-prepared for the uncertainties associated with climate change.

The state of the market for TR intelligence is rapidly evolving. Many analytics platforms and data sources are becoming available for the first time, creating a dynamic landscape where companies are beginning to experiment with different products. This shake-out period is essential as organizations test these tools to determine which questions need answering, which questions can be answered, and which tools are most effective in providing those answers. Among these opportunities, low-cost scenario scoping stands out as a crucial development. This approach allows companies to quantify the likely range of possible outcomes under different climate scenarios, helping them to better understand potential risks and opportunities.

Stress testing is another significant application of climate intelligence. This process involves evaluating how different climate-related scenarios might affect a company's financial health. For instance, banks are using stress testing to evaluate the resilience of corporate credit portfolios to transition related risks and opportunities. The insights gained from these tests can inform strategic adjustments to mitigate potential losses and to maximize untapped opportunities.

Carbon accounting software programs have revolutionized how companies track and report their emissions. These platforms are replacing traditional Excel-based methods, offering more accurate and efficient ways to manage carbon data. Carbon accounting platforms provide comprehensive solutions that help organizations measure their carbon footprints, set reduction targets and monitor progress.

Moreover, TR intelligence is proving invaluable in due diligence processes for mergers and acquisitions (M&A) and supply chain management. During M&A due diligence, companies are increasingly incorporating climate risk assessments to identify potential liabilities and opportunities associated with climate change. Similarly, supply chain due diligence involves evaluating suppliers' climate risks to ensure that the entire value chain is resilient and sustainable. Tools from credit rating agencies provide detailed climate risk assessments that are integral to these processes.

Asset-level free cash flow (FCF) analysis is another area where climate intelligence is making a significant impact. By integrating climate risk data with financial performance metrics, organizations can better understand how climate-related factors might affect their assets' future cash flows. This analysis is crucial for making informed investment decisions and ensuring long-term financial stability.

Transition risk is a complex but essential aspect of the journey toward a sustainable future, demanding careful consideration and strategic action from businesses across all sectors. As we continue to explore the broader landscape of climate risks, we will develop an understanding of the interplay between TR, PR and Nature Risk (NR), which is crucial for developing comprehensive resilience strategies that secure both economic and environmental sustainability.

CHAPTER 4C
NATURE RISK

Summary

Nature Risk (NR) involves the potential adverse effects on organizations and economies due to the degradation of natural ecosystems and biodiversity loss. This risk arises from various factors, including habitat destruction, species extinction, and the disruption of critical ecosystem services like pollination, water purification, and climate regulation. As more than half of the global GDP, estimated at around $58 trillion, relies heavily on nature, understanding and managing NR is crucial for ensuring long-term business resilience and economic stability.

Managing NR effectively necessitates a comprehensive approach that incorporates natural capital considerations into organizational strategies. This chapter delves into the three main types of NR: physical, transition, and systemic risks, each with its distinct drivers and implications. It provides insights into how different sectors, such as agriculture, pharmaceuticals, and tourism, are impacted by NR and highlights strategies for mitigating these risks through sustainable resource management, biodiversity enhancement, and innovative practices. By aligning with regulatory frameworks and integrating nature-based solutions, organizations can mitigate NR, enhance their sustainability, and support global environmental goals.

Nature risk (NR) refers to the potential negative impacts on businesses and economies arising from the degradation of natural ecosystems and the loss of biodiversity. As of 2024, more than half of the global GDP, estimated to be around $58 trillion, is highly dependent on nature and its services. As a critical component of environmental sustainability, NR encompasses a broad range of issues, from habitat loss and species extinction to the disruption of ecosystem services such as pollination, water purification, and climate regulation. Understanding and managing NR is essential for business resilience and long-term sustainability, as the health of natural systems directly influences economic stability and societal well-being.

Nature risk encompasses various dimensions of ecological health and its interconnections with economic activities, human well-being, and societal stability. The Taskforce on Nature-related Financial Disclosures (TNFD) defines NR as comprising three primary types: physical, transition, and systemic risks, each with its own categories and impacts.

Each of the NR types encompasses distinct categories, which have their own drivers and unique implications. Physical nature risk includes both acute and chronic events that significantly impact natural ecosystems and, by extension, the organizations and economies reliant on these systems. Understanding the distinction between acute and chronic nature risks is essential for developing effective mitigation and adaptation strategies.

Acute Nature Risk: refers to sudden, short-term events that cause immediate and severe damage to ecosystems. These events are often triggered by natural disasters such as hurricanes, floods, and wildfires, which are increasingly exacerbated by climate change and human activities like deforestation and land-use changes. The 2020 Australian bushfires serve as a poignant example of acute nature risk. These fires ravaged millions of hectares of forest, caused the death of billions of animals, and significantly impacted air quality and water systems. Such catastrophic events highlight the vulnerability of natural systems to sudden shocks, which can lead to rapid declines in biodiversity and disrupt essential ecosystem services like air and water purification, carbon sequestration, and soil fertility. In the years since the bushfires, affected areas have faced prolonged ecological recovery challenges, with knock-on effects including long-term habitat loss.

The implications of acute nature risk extend beyond immediate environmental damage. They also pose significant challenges for businesses and economies dependent on stable and functioning ecosystems. For instance, sectors such as agriculture, forestry, and tourism can suffer severe economic losses due to the sudden degradation of natural resources. The increasing frequency and intensity of these acute events underscore the urgency for businesses to integrate nature risk assessments into their strategic planning and to develop robust response mechanisms to mitigate the impacts of such disasters.

Chronic Nature Risk: involves long-term, gradual changes in ecosystems that lead to persistent degradation. Unlike the abrupt nature of acute risks, chronic risks manifest over extended periods, often making them less visible but equally destructive. Key drivers of chronic nature risk include ongoing deforestation, desertification, ocean acidification, and climate change impacts such as rising temperatures and changing precipitation patterns. The continual deforestation of the Amazon rainforest exemplifies chronic nature risk. This ongoing destruction contributes to significant biodiversity loss, alters water cycles, and increases greenhouse gas emissions, thereby affecting global climate systems.

The persistent nature of chronic risks demands sustained attention and long-term strategies from businesses and policymakers. Companies in sectors like pharmaceuticals, which rely on biodiversity for natural product discovery, or agriculture, which depends on stable climatic conditions and fertile soils, face substantial threats from these slow-moving but relentless changes. Addressing chronic nature risks requires a proactive approach to conservation, sustainable resource management, and the adoption of practices that enhance ecosystem resilience. It also calls for a collaborative effort between governments, organizations, and communities to implement policies and initiatives that curb environmental degradation and promote biodiversity conservation.

Much like transition risk discussed in the previous chapter, transition nature risk categories can be broken down into several categories.

Policy and Regulatory Risk: This involves risks related to the introduction of new laws, regulations, and policies designed to protect natural ecosystems and biodiversity. These regulations might include stricter land-use laws, biodiversity conservation mandates, or bans on certain harmful activities such as deforestation and illegal wildlife trade. Businesses might face increased compliance costs, potential fines, or the need to alter their operational practices to meet new legal requirements. For example, the EU's Biodiversity Strategy for 2030 includes ambitious targets such as legally protecting a minimum of 30% of the EU's land and marine areas and restoring degraded ecosystems, which could have significant implications for industries like agriculture, forestry, and fishing.

Market Risk: This arises from shifts in market demand as consumers and investors increasingly prefer products and services that are nature-friendly and sustainable. Businesses failing to adapt to these preferences risk losing market share and investor confidence. For instance, there has been a growing trend toward sustainable and ethically sourced products, influencing market dynamics in sectors such as food and beverage, fashion, and cosmetics. Companies like Unilever and Nestlé are increasingly focusing on sustainably sourced ingredients to meet consumer demand and investor expectations.

Technological Risk: Transitioning to a nature-positive economy often requires adopting new technologies and practices that reduce environmental impact and enhance sustainability. This can involve significant capital investments and operational changes. There is also the risk of technological obsolescence, where older, less sustainable technologies become outdated. An example of technological risk can be seen in the agricultural sector, where advancements in sustainable farming techniques, such as precision agriculture and agroforestry, are required to reduce the ecological footprint of food production.

Reputational Risk: Businesses face reputational risks if they fail to align their operations with societal expectations regarding nature conservation and biodiversity. Public awareness and concern about environmental degradation are growing, and companies perceived as contributing to the destruction of natural ecosystems may suffer damage to their brand and consumer trust. For example, companies implicated in deforestation, such as palm oil producers, have faced significant backlash from consumers and activists, leading to calls for more sustainable practices and transparency in supply chains.

Operational Risk: Transition nature risk also includes the operational challenges associated with shifting toward sustainable practices. This can involve disruptions in supply chains, increased costs of raw materials sourced sustainably, and the need to retrain employees or restructure business models. For instance, the fashion industry is experiencing operational risks as it moves toward more sustainable sourcing of materials and ethical production processes, which often require significant changes in traditional manufacturing processes.

Systemic nature risks are characterized by their complexity and the far-reaching consequences they can have on interconnected systems. Understanding and managing these risks is essential for building resilience and ensuring sustainable development.

Economic and Financial Stability: The degradation of natural ecosystems can undermine the stability of economies and financial systems. For instance, the loss of pollinators can severely impact agricultural yields, leading to food shortages and increased prices, which in turn can cause economic instability and social unrest. The collapse of fisheries due to overfishing and habitat destruction can devastate local economies that depend on fishing as a primary source of income and employment. These disruptions can cascade through the economy, affecting related industries and financial markets.

Health and Social Well-being: The decline in ecosystem health can directly impact human health and social stability. For example, deforestation and habitat destruction can increase the incidence of zoonotic diseases, as humans and wildlife come into closer contact.

The COVID-19 pandemic, which is believed to have originated from wildlife, underscores the profound impacts that ecological degradation can have on global health and economies. Furthermore, the loss of natural resources can exacerbate poverty and inequality, as communities that rely on these resources for their livelihoods face increased hardship.

Climate Regulation and Natural Disasters: Healthy ecosystems play a critical role in regulating the climate and mitigating the impacts of natural disasters. Forests and wetlands, for example, act as carbon sinks, sequestering carbon dioxide and helping to regulate global temperatures. The destruction of these ecosystems can accelerate climate change, leading to more frequent and severe natural disasters such as hurricanes, floods, and droughts. These events can cause widespread damage to infrastructure, displace populations, and strain government resources.

Global Supply Chains: The interconnected nature of global supply chains means that disruptions in one part of the world can have ripple effects across the globe. For example, the deforestation of the Amazon rainforest can impact global weather patterns, affecting agricultural production in distant regions. Similarly, the decline of fisheries can disrupt seafood supply chains, affecting food security and trade. Businesses that rely on global supply chains must consider the systemic nature risks that could disrupt their operations and affect their bottom line.

Policy and Governance: Systemic nature risks also pose challenges for policymakers and governments. The complexity and interconnectivity of these risks require coordinated and comprehensive policy responses that address the root causes of environmental degradation. Governments must work together to implement policies that promote sustainable resource management, protect biodiversity, and mitigate climate change. Failure to address systemic nature risks can lead to policy failures, increased conflict over natural resources, and governance challenges.

Understanding systemic nature risk requires a holistic approach that considers the interdependencies between natural ecosystems, human societies, and economic systems. By recognizing the broader implications of environmental degradation, organizations, governments, and communities can develop strategies to mitigate these risks and build resilience. This includes investing in nature-based solutions, enhancing ecosystem conservation efforts, and promoting sustainable development practices that protect both people and the planet.

Understanding the impact of NR on organizations and the economy requires a clear definition of natural capital. Natural capital refers to the world's stocks of natural assets, including geology, soil, air, water, and all living things. These natural resources provide ecosystem services that are essential for human survival and economic activity, such as pollination of crops, purification of air and water, and climate regulation partly through the removal of CO_2. Economically quantifying natural capital involves assessing the value of these ecosystem services and integrating them into financial and strategic decision making processes.

The degradation of natural capital directly affects organizations and economies by disrupting the services that ecosystems provide. For example, the decline in pollinator populations due to habitat loss and pesticide use can lead to reduced agricultural yields, impacting the food industry and increasing prices for consumers. Similarly, deforestation can lead to soil erosion, reducing land productivity and affecting sectors like agriculture and forestry. Indirect impacts arise from the broader economic and social consequences of ecosystem degradation. For instance, water scarcity caused by over-extraction and pollution can limit industrial activities, increase costs for water-dependent processes, and trigger conflicts over water resources and access. This is especially concerning when foreign ownership of key resources, like the 11.3% of Australian water entitlements held by foreign interests, is on the rise.

Agriculture relies heavily on natural capital for fertile soil, water, and pollination, all of which are under increasing stress from climate change. The decline in soil health due to unsustainable farming practices has led to reduced crop yields and increased costs for fertilizers and soil restoration. The depletion of soil organic matter in intensive farming regions has also necessitated significant investment in soil conservation measures. The challenge of meeting a doubled food demand by 2050 is further intensified by the vulnerability of crops; many crops, like coffee, are highly sensitive to even minor temperature fluctuations. This sensitivity, coupled with the aforementioned compounding stress factors of climate change, paints a concerning picture for future food security.

The pharmaceutical industry depends on biodiversity for the discovery of new drugs. The loss of biodiversity in tropical rainforests, which are hotspots for medicinal plants, can limit the availability of new compounds for drug development and can slow down the discovery of new treatments and increase the costs of research and development. Tourism is another sector heavily dependent on natural capital. Destinations that offer pristine natural environments, such as coral reefs, forests, and wildlife reserves, attract millions of visitors each year. The degradation of these natural assets due to climate change, pollution, and over-tourism can reduce the attractiveness of these destinations, leading to declines in tourism revenue. The Great Barrier Reef, for example, has suffered from coral bleaching due to rising sea temperatures and ocean acidity, impacting the tourism industry in Australia.

Crucially, the insurance industry is increasingly recognizing the financial risks associated with natural capital degradation. Insurers such as Swiss Re are incorporating environmental factors such as flood protection provided by wetlands and forests into their risk assessments to better understand and price the risks related to natural disasters, biodiversity loss, and ecosystem decline.

Effectively managing and mitigating nature risk is essential for organizations to ensure long-term sustainability and resilience. This involves integrating natural capital considerations into business strategies, recognizing the value of natural resources, and understanding the dependencies and impacts on ecosystems. Organizations can adopt a multi-faceted approach that includes sustainable resource management, enhancing biodiversity, and leveraging innovative practices to reduce their ecological footprint. This might include the adoption of renewable energy sources, such as solar or wind power, to decrease reliance on fossil fuels and reduce greenhouse gas emissions. Energy efficiency measures, such as upgrading to energy-efficient appliances and optimizing industrial processes, can also play a significant role. Additionally, organizations can focus on waste reduction through recycling programs, minimizing single-use plastics, and promoting circular economy principles, which aim to keep resources in use for as long as possible through reuse, repair, and recycling.

Enhancing biodiversity is a critical component of managing NR. Organizations can support biodiversity through various initiatives, such as establishing conservation areas and protecting natural habitats from degradation. Restoration projects are equally important; they aim to rehabilitate degraded lands, such as reforesting areas that have been deforested or restoring wetlands that have been drained. Supporting biodiversity can also be integrated into supply chain practices. Organizations can ensure that their sourcing practices do not contribute to habitat destruction by choosing suppliers that adhere to sustainable practices such as Forestry Stewardship Council (FSC)-certified wood or Marine Stewardship Council (MSC)-certified seafood. Additionally, organizations can support initiatives that promote the use of native species in landscaping and agriculture, helping to maintain local ecosystems and providing habitats for native wildlife.

Innovative practices in sustainable resource use are essential for mitigating NR. This includes adopting new technologies and practices that reduce environmental impact and enhance sustainability. For instance, in the energy sector, the use of smart grid technology optimizes the delivery of electricity by allowing for real-time monitoring and management of energy flows, which improves efficiency and reduces waste. In manufacturing, adopting cleaner production technologies can significantly reduce pollution and resource consumption. Water recycling and closed-loop systems, for example, minimize water use and wastewater generation in industrial processes. Similarly, in construction, using sustainable building materials and incorporating green design principles can reduce the environmental footprint of new developments, as seen in the increasing use of reclaimed materials and energy-efficient building techniques.

Incorporating sustainability principles such as ease of repair and recyclability into product design has created a new market for eco-conscious consumers. By integrating these strategies into their operations, organizations can reduce their ecological footprint, enhance biodiversity, and contribute to the broader goals of environmental sustainability and resilience. These efforts not only help in conserving natural resources but also position organizations as leaders in sustainability, appealing to a growing segment of environmentally conscious consumers and investors. We will explore the competitive opportunities of such strategies later in the second part of the book.

Building on these innovative practices, leveraging nature-based solutions (NbS) presents a comprehensive approach to enhancing climate resilience. NbS harness the power of natural processes and ecosystems to mitigate the impacts of climate change while providing co-benefits such as biodiversity conservation, improved human health, and enhanced ecosystem services. NbS are defined as actions that protect, sustainably manage, and restore natural or modified ecosystems in ways that address societal challenges effectively and adaptively, providing both human well-being and biodiversity benefits. By integrating NbS into their strategies, societies can address both environmental and socio-economic challenges more effectively. For example, restoring wetlands to combat flooding and enhance water quality showcases how NbS can function. Wetlands act as natural sponges, absorbing excess rainfall and reducing the risk of floods. Additionally, they filter pollutants from water, improving freshwater resources' quality. This holistic approach not only mitigates immediate climate risks but also supports long-term environmental sustainability and resilience.

Restoring degraded ecosystems, such as reforestation and coral reef restoration, further enhances resilience against long-term climate risk by sequestering carbon and protecting biodiversity. We live in a time where the Amazon rainforest, once a vast carbon sink absorbing massive amounts of CO_2, now emits more carbon dioxide than it can capture due to fires and deforestation. Damaged ecosystems such as this have the power to precipitate negative feedback processes, which can actually accelerate climate change if no remedial action is taken. However, the opposite is also true for proactive restoration. For instance, the Great Green Wall initiative aims to restore 100 million hectares of degraded land across the Sahel region by 2030, combatting desertification and promoting food security. Natural infrastructure solutions like green roofs, urban forests, and permeable pavements enhance urban resilience by reducing heat islands, improving air quality, and managing stormwater. Singapore's "City in a Garden" vision incorporates these elements to create a sustainable and livable urban environment. Coastal ecosystems such as mangroves and salt marshes protect shorelines from erosion and storm surges. In Indonesia, mangrove restoration has provided significant protection against coastal erosion and storm damage while supporting local fisheries.

Finally, quantifying the economic value of ecosystem services lost due to NR is also vital for understanding the full impact on organizations and economies. This valuation helps in making informed decisions about conservation and restoration efforts, and provides much-needed modern-day context to the importance of the natural world. The economic value of pollination services provided by bees, for example, is estimated to be around $235 billion to $577 billion annually. Similarly, wetlands that play a vital role in water purification, flood control, and carbon sequestration, have an estimated global economic value of $3.4 trillion per year. Biodiversity credits, a market-based approach to conservation, offer a mechanism to quantify and trade the economic value of biodiversity. These credits incentivize the preservation and restoration of natural habitats by providing a financial return for conservation efforts. Integrating biodiversity credits into economic valuations underscores the tangible benefits of protecting and enhancing ecosystem services, thereby aligning economic interests with environmental sustainability.

While innovative practices and strategic management are crucial for mitigating NR, organizations must also navigate a complex landscape of regulatory compliance to ensure their efforts align with broader environmental objectives. Effective NR management is intrinsically linked to understanding and adhering to global and national regulations that govern biodiversity conservation and ecosystem protection. By aligning their strategies with these regulatory frameworks, organizations can not only mitigate risks but also seize opportunities to contribute to global sustainability goals and enhance their resilience. This transition highlights the importance of regulatory compliance as both a challenge and an avenue for creating value through sustainability initiatives. Compliance with NR regulations helps organizations mitigate the adverse impacts of their activities on natural ecosystems, thereby reducing nature risk. Global and national regulations, such as the Kunming-Montreal Global Biodiversity Framework (GBF), Convention on Biological Diversity (CBD), the EU's Biodiversity Strategy for 2030, and the U.S. Endangered Species Act, set stringent standards for conservation efforts. These regulations require organizations to adopt practices that minimize harm to ecosystems and promote biodiversity.

Holistic Risk Consideration

The interplay between PR, TR and NR demands a nuanced approach to climate risk management. As organizations look toward decarbonization and climate adaptation, understanding these risks in isolation will prove insufficient due to their interdependent nature.

Decisions around physical infrastructure, for example, such as the construction of flood defenses, must take into account transition risks like future carbon taxation or commodity pricing that might render such assets obsolete. Similarly, investments in electrification, aimed at reducing carbon footprints, must also consider the enhanced physical risks of system failures without adequate hardening against climate impacts. Nature-based solutions, like carbon offsets from forestry, must be scrutinized for their vulnerability to wildfires and droughts, highlighting the cross-cutting implications of nature-related risks.

The necessity of viewing these risks through an integrated lens is also crucial for effective capital allocation and risk management ensuring that investment decisions are not only sound in the short term but sustainable in the long term, addressing the full spectrum of risks and their potential cascading effects. Understanding the interconnected exposures of PR, TR and NR over time allows organizations to craft strategies that are both resilient and adaptive, ensuring they remain viable as climate conditions evolve.

A comprehensive approach to climate risk not only aids in identifying potential financial impacts within asset cash flows but also ensures that materiality thresholds – critical for prudent investment and operational decisions – are adequately considered. This integrated perspective forms the bedrock of effective climate governance, enabling businesses to navigate the complexities of a changing climate landscape, make better decisions and align their operations with sustainable, future-proofed strategies.

This chapter serves as a foundational guide, illustrating the vast array of climate risks and opportunities, urging stakeholders to adopt an integrated approach. By understanding the synergistic and sometimes antagonistic relationships between different types of climate risks, organizations and policymakers can devise more robust strategies. Such integrated risk management not only mitigates adverse effects but also leverages potential opportunities for resilience and sustainability.

CHAPTER 5
THE POLICY RESPONSE TO CLIMATE CHANGE

Summary

This chapter explores the intricate landscape of climate policy and its crucial role in driving climate action and shaping economic and social structures toward a resilient, low-carbon future. We trace the evolution of climate policy from pioneering international agreements such as the Paris Agreement to contemporary regulation frameworks, highlighting the significance of these policies in mitigating greenhouse gas emissions and preparing for the inevitable impacts of climate change through adaptation strategies.

We examine global collaborative efforts and national policy frameworks, showcasing diverse approaches to carbon pricing, renewable energy mandates and energy efficiency standards as well as city-level and sectoral initiatives, demonstrating their pivotal role in complementing national efforts and addressing specific industry challenges.

Finally, we delve into the complexities of policy implementation, including the standardization and comparability of reporting frameworks alongside the integration of nature and climate risk in the next generation of disclosures. We discuss how climate policy can act as a catalyst for change, driving investment in clean technologies and fostering economic resilience.

Climate policy encompasses the measures and strategies implemented by governments and organizations worldwide to address and mitigate the impacts of climate change. These policies play a crucial role in driving climate action, shaping economic and social structures and influencing the global transition to a resilient, low-carbon economy. The growing urgency for effective climate policies is underscored by the increasing severity of climate change impacts, which threaten ecosystems, economies and communities worldwide.

From the early international agreements in the 1990s, through the landmark Paris Agreement of 2015, and onward to contemporary disclosure practices, climate policy has evolved significantly. These policies encompass a wide range of measures, including carbon pricing, renewable energy mandates, energy efficiency standards and adaptation strategies aimed at enhancing resilience to climate impacts. By integrating mitigation and adaptation, climate policies strive to reduce greenhouse gas emissions while also preparing for and responding to the consequences of a changing climate.

Understanding climate policy has become essential for organizations as it influences regulatory landscapes, drives technological advancements and shapes market dynamics. Effective governance, integration of climate considerations into corporate strategies and proactive engagement in policy development are critical for navigating the transition to a low-carbon and climate-resilient economy.

A growing body of scientific evidence stimulated early trans-boundary conversations on climate in the 1970s, progressing under the UN to the first Earth Summit in Rio de Janeiro during 1992. At this conference, the UN Framework Convention on Climate Change (UNFCCC) was agreed, paving the way for a more structured policy response to the emerging climate crisis. The implementation of key international agreements in the 1990s marked a significant turning point in global climate policy, most notably the Kyoto Protocol, adopted in 1997.

The Kyoto Protocol represented a groundbreaking effort by the international community to address climate change through legally binding commitments. Under this treaty, developed countries committed to reducing their greenhouse gas emissions by an average of 5% below 1990 levels during the first commitment period (2008–2012). The protocol also introduced three innovative market-based mechanisms to facilitate emissions reductions: International Emissions Trading (IET), the Clean Development Mechanism (CDM) and Joint Implementation (JI). JI allowed industrialized countries to invest in emission reduction projects in other industrialized countries, while CDM facilitated such investments in developing nations, promoting sustainable development. IET enabled countries to trade emission allowances to meet their targets.

Despite its pioneering nature, the Kyoto Protocol faced several obstacles. Binding targets were only imposed on developed countries and not rapidly industrializing nations like China and India, which were significant and growing sources of emissions at the time. This led to questions on the overall effectiveness of the protocol in addressing global emissions. The withdrawal of key countries, most notably the US in 2001, undermined the protocol's potential impact. The lack of stringent enforcement mechanisms further hindered its success, as countries failing to meet their targets faced limited consequences.

Regardless, the Kyoto Protocol played a crucial role in raising awareness about climate change and setting the stage for future international agreements by demonstrating the feasibility of international cooperation on climate issues. The protocol's emphasis on market mechanisms and its framework for emission reduction targets laid the groundwork for more inclusive and flexible approaches to global climate governance, ultimately influencing the design of contemporary climate policies at international, national and regional levels.

The European Union Emissions Trading System (EU ETS), launched in 2005, was inspired by the Kyoto Protocol's emissions trading mechanism. The EU ETS became a cornerstone of the EU's climate policy, setting a cap on the total amount of greenhouse gases that can be emitted by installations covered by the system and allowing companies to buy and sell emission allowances as needed. Although the US did not ratify the Kyoto

Protocol, the agreement spurred sub-national and state-level initiatives aimed at reducing greenhouse gas emissions. States like California took the lead in implementing ambitious climate policies, such as the Global Warming Solutions Act (AB 32) in 2006, which aimed to reduce California's greenhouse gas emissions to 1990 levels by 2020. Additionally, regional initiatives like the Regional Greenhouse Gas Initiative (RGGI) in the northeastern US created a cooperative effort among states to cap and reduce CO_2 emissions from the power sector.

Building on the momentum generated by these earlier efforts, the Paris Agreement was adopted in 2015. It remains one of the most significant milestones in global climate governance, establishing a comprehensive framework for international cooperation to combat climate change. At its core, the agreement aims to limit global temperature increases to well below 2°C above pre-industrial levels, with efforts to limit the rise to a key scientific threshold of 1.5°C. While the importance of half a degree more temperature rise is often met with confusion, the social, environmental and financial implications between both scenarios cannot be overstated. At 2°C, the global population percentage exposed to extreme heat waves will double to nearly 40% (Buis, 2019), more than 99% of coral reefs will be lost and global GDP losses will double by 2100 (Levin, 2018).

As part of the Paris Agreement, countries are required to submit their own emission reduction plans, known as Nationally Determined Contributions (NDCs). These NDCs are critical as they collectively represent the global effort to curb greenhouse gas emissions. The first generation of NDCs, submitted in 2015, aimed for an aggregate reduction in the average global temperature increase to 3.7°C. By the time of the second generation submissions in 2020, this target was lowered to 2.7°C. The Global Stocktake process, with its initial findings discussed at COP28 in 2023, evaluates progress toward these goals every five years, urging nations to enhance their commitments to bring the world closer to the 1.5°C target. Despite these efforts, the world is currently still on a trajectory for a temperature rise exceeding 2.5°C, indicating a significant gap in achieving the 1.5°C goal. The world's countries must still make deep cuts and take radical action to limit dangerous global warming.

The market mechanisms of the Kyoto Protocol have provided valuable lessons and frameworks for the Paris Agreement and current climate policies. IET facilitated the trading of emission allowances among countries, creating a global carbon market. The CDM, which allowed industrialized countries to invest in emission reduction projects in developing nations, has inspired similar initiatives under the Paris Agreement's Article 6. These mechanisms now include broader sustainable development goals, ensuring that projects not only reduce emissions but also contribute to the host country's environmental and social development. JI has evolved into more integrated and collaborative international climate efforts. These efforts focus on technology transfer, capacity building and financial support to enhance global climate resilience.

Challenges remain, however, particularly regarding the enforcement of NDCs and the absence of binding punitive measures for non-compliance. This continued lack of enforcement mechanisms necessitates strong political will and continuous global cooperation to ensure countries not only meet but exceed their commitments over time. The Agreement's success relies heavily on transparency, peer pressure, and the increasing ambition of successive NDCs to drive meaningful climate action.

While international frameworks like the Paris Agreement set the stage for global climate action, the translation of these commitments into actionable measures falls largely on national policy frameworks. National policies are crucial in bridging the gap between international goals and localized implementation, enabling countries to tailor their approaches based on unique economic, social and environmental contexts.

Countries around the world have adopted a variety of strategies to meet their climate commitments. These strategies often include carbon pricing mechanisms, renewable energy mandates, energy efficiency standards and comprehensive adaptation plans.

Carbon Pricing: Through mechanisms like carbon taxes and cap-and-trade systems, carbon pricing aims to internalize the external costs of carbon emissions, providing economic incentives for reducing greenhouse gas emissions. For example, in 2019 Canada implemented a nationwide carbon tax. The tax started at $20 Canadian Dollars (CAD) per tonne of CO_2 and increased by CAD$10 per year until it reached CAD$50 per tonne in 2022. This increase continues annually by CAD$15 per tonne from 2023 onward until it reaches CAD$170 per tonne by 2030. The carbon tax applies to fossil fuels, with the revenue being returned to households through tax rebates, thus minimizing the economic burden on consumers. The gradual increase in carbon pricing is designed to encourage businesses and consumers to reduce their carbon footprints and invest in cleaner technologies.

Renewable Energy Mandates: These mandates require a certain percentage of energy to be generated from renewable sources. Mandates have been critical in driving investments in wind, solar and other renewable energy technologies. China's 14th Five-Year Plan for Renewable Energy, released in 2022, set ambitious targets for renewable energy use, spurring significant investments and adding 160 GW of renewable electricity capacity in the first year alone – accounting for nearly half of all global deployment. The EU is also accelerating solar PV and wind deployment in response to the energy crisis, adding a record 56 GW of solar installations in 2023, marking a 40% increase from the previous year.

Energy Efficiency Standards: New policies and targets are addressing energy efficiency standards, which are another key component of national climate policies. These standards mandate improvements in the energy efficiency of buildings, appliances and vehicles, reducing overall energy consumption and emissions. Within the transport sector, vehicles are now required to meet fuel efficiency benchmarks in many countries. Increasingly stringent building regulations now require minimum energy efficiency standards, not only in new construction but also across the rental market. Measures such as these play a crucial role in reducing energy consumption, lowering greenhouse gas emissions and achieving long-term sustainability goals.

Adaptation Strategies: These are equally important, focusing on enhancing resilience to the inevitable impacts of climate change. Countries are develop national adaptation plans to assess vulnerabilities and outline strategies to protect communities, infrastructure, and ecosystems. The Netherlands, with its long history of battling sea-level rise, has implemented extensive adaptation measures, including the construction of storm surge barriers and the reinforcement of dikes, showcasing how proactive planning can mitigate the impacts of climate change.

National policy approaches vary widely, reflecting differences in economic development, political will and societal priorities. Developed countries often have more resources and technological capabilities to implement ambitious climate policies, while developing countries may face challenges related to funding, capacity, and competing development needs. Nonetheless, the global nature of climate change necessitates a collaborative approach; wealthier nations must provide support to developing countries through financial aid, technology transfer, and capacity-building initiatives. Climate change will never remain within sovereign borders. The development and implementation of robust national policies are critical to achieving the goals set forth in international agreements like the Paris Agreement.

Expanding the policy landscape beyond national frameworks is crucial to achieving comprehensive climate action. City-level policies and sectoral initiatives have gained prominence, reflecting a more localized and industry-specific approach to addressing climate change.

City-Level Policies: Cities are on the front lines of climate change and have started to implement robust policies to reduce emissions and enhance resilience. New York City's Climate Mobilization Act aims to reduce greenhouse gas emissions from large buildings by 40% by 2030. Copenhagen has set an ambitious target to become carbon-neutral by 2025, leveraging renewable energy, energy efficiency, and sustainable urban planning.

Sectoral Initiatives: Sector-specific initiatives are essential for addressing the unique challenges and opportunities within different industries. The Network for Greening the Financial System (NGFS) is a prime example, where central banks and supervisors collaborate to enhance the financial sector's role in managing climate risks. Their efforts include integrating climate-related risks into financial stability monitoring and promoting sustainable finance. Additionally, the International Energy Agency (IEA) provides transition scenarios that guide the energy sector in reducing emissions. These scenarios outline pathways for achieving net-zero emissions, considering technological advancements and policy measures necessary for the transition.

The impact of these initiatives on specific industries is significant. For instance, the NGFS's work influences how financial institutions assess and disclose climate risks, leading to more informed investment decisions and capital allocation decisions. The IEA's transition scenarios help energy companies plan their long-term strategies, balancing the shift toward renewable energy with economic viability. These localized and sectoral efforts complement national frameworks by addressing the specific needs and circumstances of cities and industries. They foster innovation, drive targeted actions and enhance the overall effectiveness of climate policies.

Climate legislation often transcends environmental objectives, functioning as a form of industrial policy aimed at maintaining and enhancing economic competitiveness in a low-carbon future. By aligning industrial strategies with climate goals, governments can drive innovation, create jobs and ensure long-term economic resilience. The Carbon Border Adjustment Mechanism (CBAM) is a prime example of this approach. Introduced by the EU, CBAM aims to prevent carbon leakage by imposing a carbon price on imports of certain goods from countries with less stringent climate policies. This ensures that EU industries, which are subject to robust environmental regulations, are not at a competitive disadvantage compared to foreign producers who may not face similar carbon costs.

CBAM is designed to level the playing field, incentivizing global industries to adopt cleaner technologies and reduce emissions. By making the cost of carbon explicit in the price of goods, it encourages innovation in low-carbon technologies and processes, not just within the EU, but also internationally as trading partners adjust to the new economic landscape. Implemented gradually starting in 2023, with full operationalization by 2026, CBAM covers sectors like cement, iron and steel, aluminium, fertilizers and electricity. For instance, a steel manufacturer in the EU must adhere to strict carbon emission regulations, increasing production costs. Without CBAM, cheaper, higher-emission steel from non-EU countries could undermine the EU producer's competitiveness. CBAM addresses this by imposing a carbon price on imported steel, ensuring fair competition and incentivizing global emission reductions.

CBAM serves as a critical tool for the EU to achieve its ambitious climate goals outlined in the European Green Deal. It helps to secure a market for low-carbon products, fostering the growth of green industries and creating new economic opportunities. This integration of climate policy with industrial strategy exemplifies how governments can use legislation to drive economic transformation toward sustainability. Such mechanisms reflect a broader trend where climate policies are intertwined with economic policies, promoting sustainable development while safeguarding industrial competitiveness. This dual focus ensures that climate action contributes to economic resilience, making the transition to a low-carbon economy a driver of innovation and growth rather than a cost burden.

Implementing climate policy effectively involves navigating numerous challenges and nuances, particularly in standardization and comparability. The development and implementation of robust climate policy standards are crucial for ensuring that efforts are coherent, transparent and effective across different jurisdictions and sectors. However, the landscape of climate reporting frameworks is complex and evolving. Transparent and standardized climate-related financial disclosures have become paramount for businesses, investors and regulators in recent years. Various frameworks have emerged to guide organizations in reporting their climate-related risks, opportunities and impacts, reflecting the growing importance of these disclosures for consistent and comparable reporting across organizations and countries.

Existing Reporting Frameworks

The Task Force on Climate-related Financial Disclosures (TCFD) was established in 2015 to develop voluntary, consistent climate-related financial risk disclosures for companies, helping investors, lenders and insurers make informed decisions. TCFD's recommendations focus on governance, strategy, risk management and metrics, providing a comprehensive approach to climate risk reporting. In 2023, TCFD transitioned its responsibilities to the International Sustainability Standards Board (ISSB), marking the culmination of its mission to mainstream climate-related financial disclosures.

The Global Reporting Initiative (GRI) has been a pioneer in sustainability reporting standards, offering a broad framework for ESG disclosures. GRI's standards cover a wide range of topics, including environmental impact, social responsibility and governance practices. Recent updates have expanded GRI's focus to include biodiversity and ecosystem impacts, ensuring a more holistic approach to sustainability reporting.

The Carbon Disclosure Project (CDP) plays a critical role in environmental data collection and disclosure. Through its comprehensive questionnaires on climate change, water and forests, CDP enables companies to disclose their environmental impacts transparently. These disclosures are crucial for investors seeking to evaluate corporate environmental performance and drive more sustainable investment practices.

The Sustainability Accounting Standards Board (SASB) provides industry-specific standards for ESG reporting, emphasizing financial materiality and investor relevance. SASB's standards help companies identify and disclose financially material sustainability information that is crucial for investors.

Next-Generation Reporting Frameworks

As the landscape of climate policy continues to evolve, next-generation reporting frameworks are increasingly emphasizing the importance of financial materiality and the assessment of opportunities for value creation, ensuring that sustainability efforts are integrated into core business strategies.

The ISSB, established as part of The International Financial Reporting Standards (IFRS) Foundation and formed as a successor to TCFD, aims to create a comprehensive global baseline for sustainability reporting. ISSB integrates TCFD's recommendations into the IFRS sustainability disclosure standards, ensuring continuity and enhancing the global standardization of sustainability disclosures. This initiative is expected to significantly impact global sustainability reporting, promoting consistency and comparability across regions.

The European Sustainability Reporting Standards (ESRS), developed under the Corporate Sustainability Reporting Directive (CSRD), mandate comprehensive sustainability reporting for companies within the EU. These standards align with global frameworks and require detailed disclosures on a wide range of sustainability topics, reinforcing the EU's leadership in sustainability governance.

The Science Based Targets for Nature (SBTN) initiative focuses on setting science-based targets for nature, addressing critical areas such as water, land, biodiversity, and oceans. SBTN builds on the success of the Science Based Targets initiative (SBTi) for climate, providing a structured approach for companies to set and achieve ambitious nature-related goals.

The Natural Capital Protocol offers a standardized framework for measuring and valuing impacts and dependencies on natural capital. This protocol helps companies integrate natural capital considerations into their decision making processes, promoting more sustainable business practices.

Nature-Related Reporting Frameworks

The Taskforce on Nature-related Financial Disclosures (TNFD) was established to develop a framework for nature-related risks and opportunities, aligning with TCFD principles to ensure integrated reporting. TNFD aims to enhance the understanding and management of nature-related risks, fostering better decision making in financial and corporate sectors.

The Global Biodiversity Framework, adopted under the Convention on Biological Diversity (CBD), sets strategic goals for biodiversity conservation through 2030. This framework emphasizes the integration of biodiversity considerations across sectors, highlighting the importance of protecting natural ecosystems in achieving sustainability goals.

Implementing climate policy effectively involves navigating numerous challenges and nuances, particularly in standardization and comparability. The development and implementation of robust climate policy standards are crucial for ensuring that efforts are coherent, transparent and effective across different jurisdictions and sectors. However, the landscape of climate reporting frameworks is complex and evolving. Transparent and standardized climate-related financial disclosures have become paramount for businesses, investors and regulators in recent years. Various frameworks have emerged to guide organizations in reporting their climate-related risks, opportunities and impacts, reflecting the growing importance of these disclosures for consistent and comparable reporting across organizations and countries.

One of the primary challenges in climate policy implementation is the standardization and comparability of reporting frameworks. Variations in frameworks like the TCFD can lead to inconsistencies, making it difficult for stakeholders to compare company disclosures effectively. For example, a financial institution might report its climate risks and opportunities using TCFD guidelines, while another might use a different set of standards. This variation can create challenges for investors and regulators attempting to assess and compare the climate-related financial risks of these institutions. Efforts to harmonize standards, such as the integration of TCFD into the ISSB, aim to address these challenges and improve the comparability of sustainability reports. The ISSB's consolidation of TCFD's recommendations into its broader framework promises a more consistent approach, which will enhance the reliability and comparability of climate disclosures across different companies and industries.

Another critical aspect is the integration of climate and nature reporting. As companies are increasingly required to disclose a wide range of environmental metrics, the challenge lies in aligning these diverse metrics into a unified reporting framework. For instance, a global consumer goods organization may need to report on both its greenhouse gas emissions and its impact on biodiversity. Historically, these disclosures might have been reported separately, creating a fragmented picture of the company's overall environmental impact. The TNFD aims to integrate nature-related risks into corporate reporting, complementing existing climate-focused frameworks. By aligning its reporting with both TNFD and ISSB guidelines, the organization can provide comprehensive and coherent sustainability reports, enhancing transparency and accountability. This integration ensures that stakeholders receive a holistic view of a company's environmental impact, facilitating better decision making and driving more effective climate and nature conservation efforts.

Climate Policy as a Catalyst for Change: The Inflation Reduction Act (IRA) – USA, 2022

Climate policy often serves as a powerful catalyst for change, driving investment in clean technologies and refocusing corporate strategies toward sustainability. The speed and depth of regulation can significantly impact how businesses respond to climate challenges. For instance, the Inflation Reduction Act (IRA) in the United States has provided substantial funding and incentives for renewable energy projects, electric vehicles (EVs), and energy efficiency improvements. This legislation has catalyzed significant private sector investment, accelerating the transition to a low-carbon economy.

The IRA, enacted in 2022, represents a landmark legislative effort by the United States to address climate change through substantial investment and incentives. This comprehensive act aims to accelerate the transition

(*continued*)

to a low-carbon economy by funding renewable energy projects, promoting energy efficiency and encouraging the adoption of EVs.

Renewable Energy Investments: The IRA allocates more than $370 billion for renewable energy projects, including solar, wind, and geothermal power. This funding is designed to support the development of clean energy infrastructure and reduce reliance on fossil fuels. Substantial tax credits are provided for solar and wind energy installations, significantly lowering the cost of these projects for developers. As of 2024, these incentives have contributed to a 40% increase in new renewable energy installations compared to pre-IRA levels.

Electric Vehicle Incentives: To promote the adoption of EVs, the IRA includes incentives such as tax credits for consumers purchasing electric cars. These credits, which can be up to $7,500 per vehicle, are intended to make EVs more affordable and competitive with traditional gasoline-powered vehicles. Additionally, the Act funds the expansion of EV charging infrastructure, addressing a critical barrier to widespread EV adoption. By 2024, EV sales have surged by 35%, and the number of public charging stations has doubled.

Energy Efficiency Improvements: The IRA also focuses on enhancing energy efficiency across various sectors. It provides funding for retrofitting buildings with energy-efficient technologies, reducing energy consumption and lowering greenhouse gas emissions. Grants and loans are available for homeowners and businesses to upgrade insulation, install energy-efficient windows and adopt smart energy management systems. As a result, energy consumption in commercial buildings has decreased by 15% since the Act's implementation.

The IRA has already catalyzed significant private sector investment in clean technologies. For example, following the act's passage, several major companies announced large-scale renewable energy projects, leveraging the tax credits and funding provided by the IRA. This influx of investment is expected to create thousands of jobs in the clean energy sector, contributing to economic growth while advancing climate goals. The incentives for EV adoption have led to increased sales of electric vehicles, with several automakers ramping up production to meet growing demand. The expansion of EV charging infrastructure is also progressing rapidly, making it more convenient for consumers to switch to electric vehicles.

The broader impact of the IRA within the global climate policy landscape cannot be overstated. By providing substantial financial incentives and setting a robust regulatory framework, the IRA has positioned the United States as a leader in climate action. This approach serves as a model for other nations, demonstrating how targeted policy measures can drive significant progress in reducing emissions and promoting sustainable development. While the IRA represents a significant step forward, its success depends on effective implementation and ongoing support. Challenges include ensuring equitable access to the benefits of the act, particularly for low-income communities, and addressing supply chain constraints for renewable energy components and EV batteries.

Looking ahead, the IRA is expected to play a critical role in reducing the United States' greenhouse gas emissions and positioning the country as a leader in the global transition to a sustainable economy. Its emphasis on investment and incentives serves as a model for other nations seeking to drive climate action through policy measures to ensure competitiveness.

References

Buis, A. (2019, 19 June). A Degree of Concern: Why Global Temperatures Matter. NASA's Global Climate Change website. https://climate.nasa.gov/news/2865/a-degree-of-concern-why-global-temperatures-matter/#:~:text=At%201.5%20degrees%20Celsius%20warming,at%201.5%20degrees%20Celsius%20warming.

Levin, K. (2018, 7 October). Half a Degree and a World Apart: The Difference in Climate Impacts Between 1.5°C and 2°C of Warming. World Resources Institute. https://www.wri.org/insights/half-degree-and-world-apart-difference-climate-impacts-between-15c-and-2c-warming

CHAPTER 6
THE CLIMATE INTELLIGENCE REVOLUTION

Summary

We explore the transformation of climate intelligence from an initial focus on carbon metrics to embracing a broad spectrum of climate-related risks and opportunities – Unified Climate Intelligence (UCI). This journey includes understanding three of the fundamental pillars of UCI: accessibility, decision usefulness and digitalization, as well as the instrumental role of advancements in artificial intelligence (AI), machine learning (ML), generative AI (Gen-AI) and big data analytics in enhancing the accuracy, depth and accessibility of climate insights. Such technological enablers have not only democratized access to complex climate data but also facilitated the integration of environmental, financial and social data, marking a significant shift toward a more comprehensive approach to climate action.

UCI's scalability across different organizational and regional levels – from the enterprise as a whole to specific assets – underscores its utility in making climate risk a "knowable" and manageable factor for decision makers. This adaptability is crucial for navigating the global transition to a climate economy, where economic activities are aligned with the objectives of sustainability and resilience against climate change.

Climate intelligence (CI) is not just a tool for compliance or risk management; it's a strategic imperative that is reshaping how businesses operate, innovate and create long-term value in a world increasingly shaped by climate change. It is a transformative force that promises to revolutionize how we navigate the challenges and capitalize on the opportunities; an agent capable of delivering change on the global scale and at the urgent pace required in the face of the climate crisis.

The evolution of CI has been a journey of increasing sophistication and depth of understanding, paralleling advances in technology, data science and computational methods. An awareness of the magnitude of change that Earth's climate could experience existed in the nineteenth century when the Ice Age Theory was proposed, most notably by Louis Agassiz. Following this, the twentieth century brought the development of the first numerical weather prediction models. Advances in atmospheric science, combined with the development of computers in the mid-twentieth century, further allowed for the creation of the first general circulation models (GCMs) of the atmosphere. These models started to incorporate ocean dynamics, recognizing the integral role of oceans in climate systems. Finally, the launch of the first weather satellites in the late twentieth century provided a global perspective on weather and climate, revolutionizing data collection and leading to significant improvements in forecasting abilities.

The twenty-first century has seen significant advancements in climate science and technologies, with the IPCC playing a pivotal role in synthesizing scientific research and disseminating knowledge about climate change to formulate global and national policy. This period marked the beginning of a more structured understanding of climate dynamics and their implications for various sectors and geographies, though still very much geared toward policymakers and scientists as a primary audience. A continued proliferation of computing power and the advent of big data analytics has led to increasingly sophisticated climate models that now incorporate a wide range of variables, including anthropogenic activities and greenhouse gas (GHG) emissions alongside increasing global temperatures and other environmental data. These models have improved in accuracy and resolution, providing detailed projections of future climate scenarios and ultimately leading to a growing recognition of climate change as a significant risk factor for businesses, governments and communities worldwide.

The integration of novel technologies, such as artificial intelligence (AI) and machine learning (ML), has revolutionized CI over the past decade transforming the way we understand and respond to climate change. They have enabled the collection, analysis and interpretation of vast amounts of climate data into actionable, practical and accessible insights, allowing for more accurate predictions and informed decision making. The fusion of environmental, financial and social data marks a significant shift from the era of expertise-specific silos to a more integrated and accessible approach to climate action. Today, CI platforms leverage these advanced technologies to provide tailored insights to specific assets, sectors and regions alike. They are becoming essential tools for businesses, policymakers and organizations seeking to understand, capitalize and manage climate risk and opportunity.

Concurrent with technological developments, CI has also evolved over the last few years from a primary focus on carbon metrics to a broader evaluation of comprehensive physical risk, to a more holistic approach integrating nature and transition risk as well, which we define as Unified Climate Intelligence (UCI). As covered previously in Chapter 3, addressing climate risk and opportunity effectively requires a 360° view that encompasses not just carbon emissions but also other GHGs, biodiversity, water usage, social equity and economic sustainability. The rise of CI and further development of UCI is in part recognition of and in service to the multifaceted solution required to the direct *and* indirect aspects of our global climate challenge. This evolution mirrors the emerging concept of the "climate economy" or "climate-aligned" economy, reflecting a journey toward a target state of affairs that not only mitigates climate risks but also aligns economic activities with the objectives of climate sustainability and resilience.

For enterprises particularly, the challenge of climate-alignment is compounded by constraints such as limited budgets, a lack of in-house expertise and the need to navigate complex regulatory environments. Investment and divestment decisions cannot merely be based on historical metrics that do not factor in future forces that are shaping the economy; instead, they must adapt to evolving goalposts of longer-term climate value creation. These shifting standards necessitate interventions that may entail substantial upfront costs but are crucial for ensuring long-term viability. Such interventions encompass a broad spectrum of actions, including immediate reduction and mitigation efforts, as well as ongoing adaptation strategies to align with future sustainability goals. As such, UCI provides a strategic framework for enterprises that are increasingly called upon to become more climate-conscious, to transition toward a green economy, all while retaining profitability and growth.

UCI can be defined as a comprehensive framework that assesses and integrates a wide range of climate-related risks and opportunities, including physical, transition and natural capital risks, alongside social and econometric factors associated with climate change. Unlike traditional climate risk assessments that may focus on specific factors in isolation, UCI provides comprehensive, actionable insights across multiple dimensions of climate change. This includes incorporating the Shared Socioeconomic Pathways (SSPs) and scenarios set by the IPCC, adapted scenarios from Integrated Assessment Models as well as specialist datasets covering physical, transition and nature related drivers (e.g., energy and resource use, water use, species datasets, land use change and much more). By synthesizing data from diverse sources and considering scenarios ranging from high-emission pathways to those aligned with Paris Agreement targets, UCI provides stakeholders with actionable insights that reflect the complex interactions and feedback loops between environmental, economic and societal systems. This holistic approach enables policymakers, businesses, investors and communities to develop robust strategies and policies that enhance resilience, promote sustainable growth and facilitate effective and future-proofed climate action.

Three of the fundamental pillars of UCI ensure its effectiveness across various sectors: accessibility, decision usefulness and digitalization. Accessibility ensures that UCI's comprehensive climate insights are readily available to all stakeholders, regardless of their technical background, enabling informed decision making across the board. Decision usefulness guarantees that the insights provided are not only relevant but also actionable, helping users from businesses to policymakers to make strategic choices in response to climate risks and opportunities. Finally, digitalization is the backbone of UCI, leveraging the latest in technology to analyze vast datasets, offering precise and up-to-date climate intelligence that can drive impactful climate action and sustainable development.

The accessibility of UCI is not just a matter of equity but a practical necessity for mobilizing a comprehensive global response to climate change. It is essential that the critical insights UCI can provide are available to everyone, regardless of their geographical location or financial resources. The democratization of climate action enables not just large corporations or governments but also small businesses, local communities and developing nations to make informed decisions. However, addressing the digital divide is crucial in this context, as unequal access to technology and information can hinder the ability of some regions and communities to benefit from UCI, exacerbating existing inequalities. Accessible UCI fosters collaboration across borders, sectors and disciplines. Shared, networked intelligence can lead to synergistic solutions, pooling resources, knowledge and expertise to tackle climate challenges more efficiently. Crucially, the communities most affected by climate change are often those with the least resources to combat it – challenges may affect all corners of the globe but they have disproportionate impacts on different communities. Equally, open access to climate intelligence is a critical pathway to global climate resilience, underscoring the need to bridge the digital divide and ensure equitable access to UCI and its benefits.

Making UCI accessible hinges on three key strategies: leveraging open-source and peer-reviewed climate data to ensure information is freely available and fosters transparency and innovation; employing transparent methodologies for data collection, analysis and interpretation to build trust among users; and prioritizing user-friendly interfaces that simplify complex data, making it comprehensible and actionable for non-specialists. These approaches collectively democratize climate intelligence, enabling a wide range of stakeholders to engage with and utilize climate insights. An example of this is Probable Futures, an organization that is building climate literacy by making climate data, tools, stories and resources available, accessible and relevant globally.

Decision usefulness is a fundamental principle of UCI, emphasizing the need for climate data and insights to be not just informative but actionable. UCI is structured to directly support decision making processes, offering clear, actionable guidance that enables individuals and organizations to effectively manage and respond to climate risk. UCI goes beyond data analytics; it interprets data in the context of its impact on operations, investments and policy, providing specific recommendations that can guide action. For businesses, it is important that climate science is married with enterprise specific data and core financial metrics to inform strategic and operational decisions. This might mean identifying the return on investment of measures to reduce carbon emissions or adapt to climate impacts, estimating the impact of climate change under different scenarios on future equity value or trade-offs that need to be analyzed (e.g., temperature controlling a facility reduces the exposure to future heat stress but may increase energy costs and carbon emissions). For investors, it could involve pinpointing which assets are at risk from climate change and which represent opportunities for sustainable growth. UCI can support policymakers to craft regulations and initiatives that effectively mitigate climate risks at the community or national level.

In some ways, you might think of the many applications of UCI like those of a Swiss Army knife. The key to a UCI platform's effectiveness lies in its ability to customize the presentation and application of its insights based on the specific goals, challenges and decision making contexts of its users. This might involve filtering and prioritizing information differently for a finance or sustainability executive compared to an investment manager or a city planner. Despite these different focuses, the underlying data and intelligence remain rooted in the same comprehensive understanding of climate science and impacts, ensuring that all users are drawing from a single source of truth, centered on scientific consensus.

The versatility of UCI extends to its scalability across, for example, organizational structures, from the overarching enterprise down to individual business units, supply chains or specific assets. This scalability ensures that UCI can support decision making across various levels and geographies, laying the foundation for it to be recognized as a "knowable risk." Such visibility empowers not only internal stakeholders but also external observers, who can model or assess these risks independently, thereby enhancing transparency and accountability in climate risk management.

The ability to tailor climate risk and opportunity management with the strategic priorities and operational contexts of its users, makes UCI a true twenty-first century superpower for decision makers. Insight stratification means that user requirements are inherently climate-aligned for each distinct application. For example:

Enterprise-Grade UCI could offer businesses precise recommendations on adapting operations, supply chains, and product offerings to mitigate climate risks and seize sustainability opportunities. For example, a company might use enterprise-grade UCI to decide the best locations for its new facilities to minimize exposure to physical climate risks like flooding alongside access to renewable energy, or to identify green technologies that can reduce its carbon footprint and operational costs.

Investment-Grade UCI might provide investors and financial institutions with analysis that helps them understand how climate change impacts asset values and investment risks. By incorporating UCI into financial metrics and models such as projected future cash flows or assessing Climate Value at Risk, investment-grade UCI enables a deeper insight into the climate resilience of sectors or companies. An investor could, therefore, identify which sectors or companies are best positioned to thrive in a low-carbon economy, guiding portfolio adjustments that balance financial returns with climate resilience.

Regulation-Grade UCI could assist policymakers and regulatory bodies in developing policies and initiatives that effectively address climate change at various scales. Through regulation-grade UCI, a city planner might evaluate the effectiveness of different urban green infrastructure projects in reducing urban heat islands and improving resilience to extreme weather, informing policy decisions and urban development plans.

Digitalization, through big data, machine learning and other technologies, is a pivotal force in advancing and propagating UCI. It has enhanced the accuracy, depth and breadth of climate intelligence but as discussed previously, has also democratized access to this critical information, enabling a more informed, agile, and coordinated response to the challenges posed by climate change. Central to its transformative impact is the ability to aggregate and analyze vast amounts of climate-related data from diverse sources, including satellite imagery, sensor networks and climate models. This aggregation facilitates a multidimensional view of climate dynamics, offering insights that were previously unattainable due to data volume, complexity or computational limitations. We will explore these enabling technologies further in the context of UCI architecture.

Technological enablers play a pivotal role in harnessing, processing and delivering comprehensive UCI via the integration of vast datasets, the application of advanced analytics and the provision of actionable intelligence. At the forefront of this advance in climate technology are AI, ML and big data analytics – supported by increasingly powerful cloud computing and storage, sophisticated visualization tools and exponential growth in data acquisition and integration capabilities. Crucial too are the decreasing barriers to adoption of UCI such as the falling costs associated with computing power and data storage alongside the increased access to open-source tools and the emergence of Generative AI (Gen-AI).

Big data analytics serves as the foundation that enables the effective use of ML and AI by providing the data insights necessary to train models, make predictions or automate decision processes. In recent years, big data analytics has grown more sophisticated with advancements in data storage, processing technologies and analytics algorithms, allowing for the handling of ever-larger datasets more efficiently and in real-time. This is largely due to three core dimensions of exponential technology in analytics: "compute," "data" and "algorithms." The "compute" aspect has seen a vast expansion at lower processing costs, enabling the handling of complex analytics. The "data" dimension has witnessed an explosion in the climate and environmental arena, much of it now made readily available given the climate urgency. Finally, the advancement in "algorithms" provides the edge, unlocking significant insights and enabling the synthesis of complex datasets into patterns and trends to reveal value.

By providing the groundwork to understand climate data, big data analytics not only enables ML and AI to interact with information in increasingly complex ways to achieve autonomous learning and intelligent decision making but also improves response times to climate-related disasters and enhances the effectiveness of mitigation and adaptation strategies. This functional immediacy, supported by the strategic deployment of Gen-AI, positions UCI as a pivotal tool in navigating the challenges and opportunities presented by climate change.

With the emergence of Gen-AI, the use and deployment of UCI across organizations will accelerate significantly. Gen-AI introduces the capability for natural language processing, allowing users at all levels to ask questions, automate workflows and generate an understanding of climate risk and opportunity. Gen-AI also significantly reduces the technical barrier needed for using UCI in day-to-day decisions and workflows. It enables users to have the power of specialized expertise such as deep climate science, policy expertise, financial analytics and more at their fingertips and all in an easy to understand and decision relevant form.

By combining advanced AI capabilities with comprehensive climate data, organizations can gain a competitive edge, accelerate decision making, and drive innovation. However, the energy consumption associated with training and running these models must be carefully considered. To fully realize the potential of AI in combating climate change, the industry must prioritize the development of more energy-efficient algorithms and hardware, coupled with a shift toward renewable energy procurement for data center operations. This dual focus on technological advancement and sustainability is crucial for ensuring that AI becomes a force for good in addressing climate change.

Finally, barriers to wider UCI adoption are beginning to decrease. The costs associated with computing power, data storage, AI and ML have notably fallen due to several key factors. Rapid technological advancements have made computing hardware more efficient and cost-effective, while the increasing demand for cloud services and processing power has led to economies of scale, reducing costs for end-users. The proliferation of open-source AI and ML frameworks has minimized software costs, facilitating the adoption of advanced algorithms without hefty investments. Cloud computing platforms like Amazon Web Services, Google Cloud and Microsoft Azure now offer scalable, affordable resources, further democratizing access to high-performance computing. Additionally, a surge in both public and private investment in AI and ML has fuelled innovation and the development of more efficient computing techniques. Together, these developments have significantly lowered the barriers to leveraging AI and ML across various fields, including climate technology, making it increasingly viable for organizations of all sizes to utilize these tools in addressing climate change challenges.

AI has expanded beyond theoretical research to practical applications, driven by the exponential increase in computational power and the availability of vast amounts of data. AI systems can now perform tasks that were previously thought to require human intelligence, from autonomous driving to personalized medicine, and have become more integrated into everyday technology. ML has seen significant progress in algorithm development, particularly in deep learning, which has led to breakthroughs in areas such as image and speech recognition, natural language processing and predictive analytics.

The application of AI and ML to climate technology has enhanced the accuracy of climate models, facilitated real-time monitoring of environmental changes, and supported the development of innovative solutions for mitigation and adaptation. Using algorithms, they can analyze historical climate data to predict future climate conditions, including extreme weather events, temperature changes and precipitation patterns.

This predictive capability is crucial for planning and preparedness in various sectors, including agriculture, water management and urban planning. ML also excels at identifying patterns within large datasets (on the order of millions of instances) that might not be apparent to traditional methods. As such, it has the power to better understand complex climate systems and identify the impacts of various climate mitigation strategies. In turn, AI enhances the accuracy of climate models, allowing for more detailed and comprehensive simulations of future climate scenarios than was previously possible. These capabilities also apply to individual companies, including optimizing energy usage or driving efficiencies that reduce the carbon footprint of operational processes or inputs.

Moreover, Gen-AI has been pivotal in broadening the accessibility and usability of UCI, making it possible for a wider audience to grasp complex climate insights without requiring the depth of expertise traditionally associated with PhD-level research. This is particularly transformative for mid to senior-level executives, serving as a prime example of how Gen-AI enables decision makers across various organizational levels to leverage UCI effectively, without the prerequisite of specialized academic credentials.

The evolution of UCI is paralleled by the dynamic and diverse ecosystem of climate technologies that aim to mitigate and adapt to climate impacts. At its foundation, climate technology spans several key clusters, which represent a foundation of interconnected solutions that are crucial for a holistic approach to climate action:

Energy Transition: Shifting from fossil fuels to renewable energy sources, including solar, wind, hydro, geothermal and emerging technologies like nuclear fusion. Energy storage, grid modernization and decentralized energy solutions are integral components of this transition.

Carbon Management: Focuses on reducing, capturing or sequestering carbon emissions. Technologies such as carbon capture and storage (CCS), direct air capture (DAC) and carbon removal methods like biochar and ocean alkalinity enhancement are key. Geothermal energy and ocean-based solutions like OTEC also contribute to carbon management efforts.

Sustainable Transportation: Encompasses the shift toward low-carbon transportation modes, including electric vehicles (EVs), autonomous vehicles, public transport and the decarbonization of aviation and shipping sectors.

Climate Adaptation: Addresses the impacts of climate change through water management, sustainable agriculture, infrastructure resilience and biodiversity conservation.

Circular Economy: Promotes waste reduction, recycling and resource efficiency through circular design, waste management and shared economy models.

Built Environment: Focuses on sustainable construction, green buildings and energy-efficient infrastructure.

Beyond these core clusters, several emerging areas of climate technology are gaining traction, expanding the frontiers of what's possible and allowing for more innovative approaches to addressing climate challenges and opportunity:

Digital Climate Solutions: Leveraging AI, machine learning, blockchain and IoT to address climate challenges.

Space-Based Solutions: Utilizing satellite technology for climate monitoring and early warning systems.

Biotechnology: Applying biological systems to develop sustainable solutions, including biofuels, bio-based materials and synthetic biology.

The interconnectedness and synergy between these clusters mean that successful implementation often requires a comprehensive, integrated approach, tailored to specific sectoral needs and global collaboration. Gen-AI powered UCI plays a pivotal role in navigating this complex landscape. For example, UCI's application in the energy sector can guide decisions on combining renewable energy sources with energy storage solutions, considering factors like geographic viability, cost implications and potential environmental impact.

A global energy company seeks to optimize its renewable energy portfolio: Traditional approaches involve extensive data analysis, complex modeling and expert judgement. By leveraging Gen-AI, the company can rapidly analyze vast datasets encompassing factors such as geographical location, weather patterns, energy demand, government policies and technology costs. Gen-AI can identify optimal combinations of renewable energy sources, storage technologies and grid infrastructure, considering factors like intermittency, energy yield and economic viability. This enables the company to make data-driven decisions about investment priorities, project siting and operational strategies.

These advanced capabilities of UCI are revolutionizing how businesses, policymakers and financial institutions engage with climate technology. By delivering insights that are not only deep and comprehensive but also directly actionable, UCI ensures that climate considerations are integral to strategic decision making processes. This holistic view, supported by fiscal incentives and regulatory frameworks such as the US Inflation Reduction Act, facilitates a more aggressive adoption of emerging technologies, stimulating innovation and driving the commercial viability of new solutions.

PART II
Building a Climate Intelligent Organization

CHAPTER 7
THE EMERGING OPPORTUNITY OF UNIFIED CLIMATE INTELLIGENCE (UCI)

Summary

In the wake of increasing climate consciousness, businesses are recognizing the necessity to integrate climate considerations into their strategic frameworks. This shift, propelled by market forces, regulatory drivers, technological advancements and radical transparency, moves beyond mere compliance, urging companies to leverage climate actions as opportunities for sustainable growth and resilience.

The evolving regulatory landscape, highlighted by initiatives such as the EU's Corporate Sustainability Reporting Directive, underscores the urgent need for businesses to adapt. Emerging regulations and alliances set a precedent for transparency and sustainability, pushing companies toward innovative solutions that align with global climate goals.

Illustrative examples from leading corporations demonstrate the tangible benefits of this strategic realignment. Companies are actively transforming their operations and business models, showcasing a commitment to sustainability that not only meets regulatory requirements but also positions them favourably in a competitive, climate-aware market.

In Part I, we covered the general foundational concepts of climate science, climate risk and opportunity and the architecture of climate technology. For the remainder of this book, we will shift focus toward the application of Unified Climate Intelligence (UCI) within the organizational context. Organizations stand at the forefront of the global response to climate change, not only because of their significant environmental footprint but also due to their potential to drive substantial positive change that is climate-aligned and reap strong financial rewards. Historically, climate considerations have lingered at the edges of corporate strategy, often seen merely through the lens of regulatory compliance, reputation management and reporting. However, the evolving climate landscape marks a permanent shift in corporate strategy, with climate considerations steadily progressing from the periphery to the core of strategic decision making. As organizations transition toward the new "climate economy," it's essential to go beyond legal minimums to unlock value, recognizing sustainability and climate action as investments for new value rather than costs. Currently though, a "Value Translation Gap" persists, especially between sustainability and finance functions, creating a disconnect between sustainability goals and resilient value creation. Put simply, organizations struggle with the ability to adequately translate climate and sustainability metrics into financial ones. This gap highlights the necessity for organizations to adapt, utilizing UCI not just for risk, regulation and reputation (something we refer to as the "3-R Mindset") but as a foundational pillar for future competitiveness and enhanced operational efficiency in a climate-conscious market. Furthermore, radical transparency will undoubtedly alter the competitive landscape, compounded by rising regulations. This demands bold leadership that transforms what is seen as climate risk into value creation opportunities.

Climate change has historically and is presently framed as a business risk – a narrative that has often led to defensive strategies focused on mitigation, reputation management and compliance. On a basic level it has been another checkbox exercise that costs time and money, but yields no tangible business return. Now however, we are already witnessing firms with a "blueprint" for the new climate economy. Take for example, trailblazing companies in the electric vehicle space, who are not only mitigating environmental risks but also capturing significant market share from traditional automotive companies. Danish energy company, Ørsted is another example of climate value creation – transitioning from fossil fuels in 2010 to become a world leader in wind energy. They are targeting 99% power generation from renewable sources by 2025 and have seen a net income flip from negative, to between approximately \$1 billion to \$3 billion from 2016 to 2022 (McKinsey & Company, 2024). They have also decreased emissions by ~90%.

Outdoor lifestyle company Patagonia has integrated climate opportunity into its branding by embodying sustainability in actions that resonate with their customers, thereby enhancing loyalty and market position. Their unique sales approaches, like the "Don't Buy This Jacket" campaign, challenge conventional consumerism by encouraging customers to reconsider consumption habits, emphasizing product durability and environmental impact. Patagonia also runs a repair and reuse program for existing clothing that not only extends the life of products but also aligns with customers' values on sustainability. These initiatives have strongly distinguished them in the marketplace, deepened customer allegiance and quadrupled sales in the last decade to hit more than \$1 billion annually.

Patagonia's strategic integration of climate and sustainability into their core mission – "We're in business to save our home planet" – exemplifies a pioneering approach to climate intelligence and circularity in the corporate world. Their commitment is reflected not only in their innovative campaigns and sustainability programs but also in fostering a culture that prioritizes the planet over profit. This ethos has not just enhanced their brand loyalty but has also set a benchmark for how businesses can thrive by aligning their operations with the urgent need for climate action. Through these endeavors, Patagonia illustrates that genuine dedication to sustainability can drive financial success while contributing positively to the planet.

Achieving climate success requires more than individual action; it demands change on a global scale. As demonstrated by industry leaders, navigating to profitability and longevity within the climate economy is feasible. And for the majority of organizations that aren't starting from scratch, success begins by comprehensively understanding your "climate health" baseline and methodically adapting with strategic, climate-aligned steps. This strategy allows businesses to evolve sustainably, emphasizing that significant, global-scale change, while challenging, is essential and achievable for long-term growth in a climate-conscious world.

Alongside integrating climate considerations into corporate strategy, there is a growing demand for businesses to be open about their environmental impacts, sustainability efforts and the impacts of climate on their current and future business performance. We are still in an era of greenwashing where phrases such as "eco-," "sustainable-" or "bio-" have lost credibility. The digital revolution has catapulted organizations into an era of radical transparency, where environmental impacts are monitored, independently measured, shared and scrutinized like never before. Simultaneously, investors are demanding greater transparency on the impacts of climate on companies' future financial performance, including actions they are taking to defend or create value in the face of powerful climate related market forces. As we transition into the climate-conscious economy, radical transparency not only meets but heightens consumer, investor and regulatory expectations. The shift is powered in part by advancements in satellite technology, AI and global data platforms that make once-hidden information readily available and actionable. The age of transparency that will unlock a powerful "I see what you see" dynamic is a permanent shift that will shape the competitiveness of organizations and shake up entire sectors.

Digitalization is propelling organizational transparency in climate efforts, where transparency fosters data sharing. However, this sharing process faces challenges due to its complexity and the lack of standardization. Technology plays a crucial role in itself overcoming these obstacles by developing frameworks that streamline and standardize the sharing of environmental data. It also has the power to support harmonization of data between organizations and national actors, or international frameworks such as the Paris Agreement, realizing the full potential of organization involvement in global climate action. This creates a virtuous cycle: transparency leads to sharing, which is facilitated by technology, further enhancing transparency.

Google Earth Engine (GEE) is a widely utilized platform that leverages powerful environmental monitoring capabilities, offering detailed observations of our planet's landscapes. It utilizes vast satellite imagery and data analysis technologies to detect and map environmental changes, such as deforestation, urban expansion and the retreat of glaciers, with remarkable precision and scale. By making complex environmental data accessible and understandable, GEE, alongside others like NASA, Microsoft and the European Space Agency, significantly contributes to enhancing transparency in climate activity. In fact, the "seed" that grew GEE was the closer investigation of a deforestation project in Santa Cruz, California by a local resident, and Google Engineer following concerns over the transparency of the operator's proposal.

The consumer and investor demand for corporate accountability has grown alongside the digital revolution and today we see many of the world's biggest organizations embedding this technology into their climate commitments and efforts to act transparently. As part of its 2050 net zero goals and Global Reforestation Project, Nestlé is utilizing high-resolution satellite imagery to monitor the health and growth of the 200 million trees it will plant by 2030. Unilever is monitoring and managing the environmental impacts of its complex global supply chain by using geospatial analytics to trace the origins of crops, identifying individual farms and plantations supplying their mills and addressing any deforestation activity associated with its production. Satellite imagery has developed to provide imagery at a staggering resolution of 30 cm and beyond land use changes, this technology is now capable of monitoring biodiversity indicators, alongside canopy density, tree height and much more.

In addition to AI/ML capabilities, blockchain is widely considered a promising new solution in enhancing organizational climate transparency (World Economic Forum, 2023). For example, these technologies offer robust solutions for tracking and managing carbon emissions effectively. Carbon credit mechanisms have come under fire repeatedly in recent times, with several top-tier organizations using credits that have been found to be effectively worthless, with the market dogged by issues like double accounting and a lack of standards and visibility. By cleaning up the transactional process and digitalizing the voluntary carbon market (VCM), blockchain provides promise in the quality and integrity of the credits to purchases. This is an essential component to successfully scaling the VCM, estimated to be worth US$50 billion by 2030.

The financial regulations surrounding climate disclosures are sharpening the focus on environmental transparency within the corporate sector. As these regulations evolve, businesses face the dual challenge of adhering to new compliance standards while seizing the opportunity to showcase their commitment to sustainability. The EU's Corporate Sustainability Reporting Directive, which expands upon the existing Non-Financial Reporting Directive, for instance, acts as a catalyst for change, pushing companies to disclose not just financial metrics but their environmental impact as well. This requirement, while challenging, opens avenues for companies to differentiate themselves, attract investment and build consumer trust by demonstrating leadership in sustainability. However, adapting to these regulatory demands poses significant hurdles. Organizations must invest in systems and processes to accurately track and report environmental data along with climate impacts on their current and future financial performance, a task complicated by the varying standards and expectations across jurisdictions.

The stakeholder shift toward sustainability is consistent across the board. Investment is growing significantly, as seen in the increasing allocation of funds toward ESG-focused projects. This shift is driven by evidence suggesting that companies with strong sustainability profiles often exhibit better operational performance and resilience to market volatility. It is also supported by initiatives like the Principles for Responsible Investment (PRI), which actively encourages investors to disclose their ESG-related activities and impacts. This transparency fosters accountability among investors and companies, leading to more informed decisions that align with long-term sustainability goals. By adhering to PRI, investors contribute to a more sustainable global financial system, driving positive change in corporate behavior through responsible investment practices.

Employees too are highly motivated by an organization's commitment to environmental and social issues; in fact, 66% of Gen Z and 57% of Millennials agree environmental concerns should take priority over economic growth Transparency is a core part of building employee trust on climate commitments and research has found that workers who trust their employers are 260% more motivated to work and 50% less likely to leave (Reichheld and Dunlop, 2023).

Consumer trends toward sustainable purchasing decisions have grown consistently year on year, with around 84% of global consumers considering sustainability important when choosing a brand (IBM 2021). Younger generations are also nearly 30% more likely to buy environmentally-conscious products. The purchasing power of Millennials and Gen Z is forecast to surpass that of Boomers around the year 2030, with a transfer of up to $68 trillion of wealth in the US alone signalling an enormous shift in consumer habits by the end of the decade (Reichheld, Peto and Ritthaler, 2023). Unilever's Sustainable Living Plan showcases strategic adaptation to meet consumer demand for sustainability, aiming for significant environmental impact reduction while enhancing social welfare. This plan, directly influenced by consumer expectations, outlines ambitious goals to improve health, reduce environmental footprint and positively impact millions of lives. It's a prime example for business leaders on how consumer-centric sustainability efforts can drive strategic innovation, operational efficiency and competitive differentiation, reinforcing the importance of integrating ethical considerations into core business practices for sustained growth and market relevance.

The regulatory landscape for climate change is rapidly evolving, with international agreements and local policies exerting significant pressure on business operations and dictating a new market dynamic. As such, businesses are adapting their operations to comply with new regulations, which often means overhauling their strategic approach to energy use, emissions and reporting as well as retraining and upskilling to address this now permanent expertise requirement. The rise of global or multi-country regulations in the climate arena, including Paris-aligned emissions targets, disclosure frameworks such as TCFD (now incorporated into IFRS) or CSRD and the Carbon Border Adjustment Mechanism (CBAM), is essential for ensuring a coordinated response to climate change. These frameworks aim to level the playing field, preventing "carbon leakage" where companies might relocate emissions-intensive activities to countries with less stringent regulations. This global regulatory architecture necessitates a strategic approach to sustainability and emissions reduction, as it directly impacts competitiveness. It has already galvanized corporate action on climate change, with two thirds of Fortune 500 firms now committed to net zero, or other significant climate targets (Climate Impact Partners, n.d.).

Regionally, approaches to climate regulation vary, with the EU currently leading in regulatory frameworks, while other regions are accelerating their policy adaptations to align with global standards. This diversity reflects the complex interplay of economic, political, and environmental factors shaping corporate responses to climate regulation globally. However, at the heart of this driving force is the requirement for companies to quantify the impact of climate on their future financial performance and to disclose actions that they plan to take to address this. Regulators, asset managers, investors, supply chain partners and other stakeholders will have the ability to assess the future climate financial performance of a company and understand their competitive positioning versus their peers. In short, climate change is an issue of financial materiality, and companies must have and disclose a strategy on how they will navigate future climate related risks and opportunities.

Regulatory measures such as the CBAM impose tariffs on carbon-intensive imports. By imposing tariffs, CBAM motivates companies to lower their carbon emissions to access markets with strict environmental standards. This dynamic not only pressures firms to adopt greener practices but also potentially favours domestic industries that already comply with such standards. In fact, we are already seeing bilateral agreements that are moving beyond simple regulation to drive green economy competitiveness such as the Australia and Singapore Green Economy Agreement (GEA) (Australian Government, n.d.). The GEA combines trade, economic and climate objectives, aiming to drive growth and opportunity for both party countries, while reducing emissions.

Investor-led initiatives to climate-align portfolios are becoming more common and directly impact organizational operations. The UN-convened Net-Zero Asset Owner Alliance (NZAOA) is a group of asset owners committed to transitioning their investments to net-zero greenhouse gas emissions by 2050. This alliance sets intermediate targets for CO_2 reduction for 2025 (22–32%) and for 2030 (40–60%), making it a pioneering force in the finance industry for setting and achieving tangible climate action goals (UN Environment Programme, n.d.). To align with the targets, businesses must implement comprehensive operational and strategic adjustments. This includes transitioning to renewable energy sources, enhancing energy efficiency across operations, investing in carbon offset projects and re-evaluating supply chains for sustainability. For example, Allianz, a member of NZAOA, is incorporating ESG criteria into all its investment decisions and aiming to be carbon neutral by 2030.

Alliances such as NZAOA influence organizations beyond their direct ownership by setting industry standards and expectations for sustainability. When large investors commit to net-zero targets, they increase pressure and drive demand for more sustainable business practices across all companies in their investment portfolios. This creates a ripple effect: companies aspiring to attract investment or maintain their attractiveness in the market must adapt by aligning with these sustainability goals, even if they're not directly owned by alliance members. In response to this market shift, sector-specific best practices are emerging. From the manufacturing sector, implementing circular economy principles, reducing waste and emissions through process optimization and sourcing sustainable materials, to the design and build of green buildings, integration of energy-efficient and low-carbon technologies and prioritization of sustainable urban planning within the construction sector, we are seeing the trickle down influence of global benchmarks from international climate frameworks into individual organizations. Approached boldly, regulation is transforming into a pathway for innovation.

At the convergence of these distinct challenges and developments, sits opportunity for action and advantage. By leveraging big data, Gen-AI and machine learning, UCI enables real-time monitoring and predictive analytics, essential for evolving regulatory compliance and strategic decision making. This approach not only caters for the rising demands of financial reporting but provides a competitive edge in the new climate economy. In the remaining chapters, we will explore UCI in action, facilitating strategic foresight and stakeholder engagement, as well as fostering a culture of climate intelligence across organizational hierarchies through transformative leadership.

References

Australian Government. (n.d.). Singapore–Australia green economy agreement. Department of Foreign Affairs and Trade. https://www.dfat.gov.au/geo/singapore/singapore-australia-green-economy-agreement

Climate Impact Partners. (n.d.). Markers of real climate action in the Fortune Global 500. https://www.climateimpact.com/news-insights/fortune-global-500-climate-commitments/

IBM. (2021, May). Sustainability at a turning point. IBM Institute for Business Value. https://www.ibm.com/downloads/cas/WLJ7LVP4

McKinsey & Company. (2024, 26 January). A different high growth story. *McKinsey Quarterly*. https://www.mckinsey.com/capabilities/strategy-and-corporate-finance/our-insights/a-different-high-growth-story-the-unique-challenges-of-climate-tech

Reichheld, A. and Dunlop, A. (2023, 31 January). Challenging the orthodoxies of brand trust. Deloitte Insights. https://www2.deloitte.com/uk/en/insights/topics/leadership/brand-trust-and-challenging-orthodoxies.html

Reichheld, A., Peto, J. and Ritthaler, C. (2023, 18 September). Research: Consumers' sustainability demands are rising. *Harvard Business Review*. https://hbr.org/2023/09/research-consumers-sustainability-demands-are-rising

UN Environment Programme. (n.d.). UN-convened Net-zero asset owner alliance. https://www.unepfi.org/net-zero-alliance/

https://www.kcl.ac.uk/policy-institute/assets/who-cares-about-climate-change.pdf

World Economic Forum. (2023, 25 April). Blockchain for scaling climate action. https://www.weforum.org/publications/blockchain-for-scaling-climate-action/

CHAPTER 8
UCI – THE NEW SUPERPOWER FOR ORGANIZATIONS

Summary

We explore the transformative superpower of Unified Climate Intelligence (UCI) as a comprehensive framework that synthesizes a wide array of climate-related risks and opportunities, integrating physical, transition and natural risks with socio-economic factors associated with climate change. We cover the concept of financial materiality and its essential integration in climate intelligence analytics, providing the financial quantification currently missing from organizational climate action.

We outline the QuantEarth™ Framework as an example of applying UCI to business operations, highlighting its capability to reduce complexities and barriers in climate-related decision making. By employing a structured three-stage process – Quantify, Plan, Act – the framework equips organizations with the tools to assess the financial impact of climate drivers, inform strategic adaptation and efficiency planning and translate insights into actionable strategies for resilience, operational efficiency and long-term value creation.

In Chapter 3, we defined Unified Climate Intelligence (UCI) as "a comprehensive framework that assesses and integrates a wide range of climate-related risks and opportunities, including physical, transition and natural capital risks, alongside social and economic factors associated with climate change." We have also highlighted the critical role that UCI plays in navigating organizations toward understanding the opportunity and risks involved in value creation and accelerating climate action – and how value will be realized by integrating UCI into core business strategies. The superpower of UCI comes from its innate transformative power to both deeply inform decision making and to enhance the certainty of these decisions.

The advancement toward UCI takes isolated, one-dimensional models and creates a cohesive, multi-faceted framework that aligns with climate goals. Combining diverse data and metrics with scientific and financial modeling, UCI transforms sustainability data into actionable insights, paving the way for strategic decisions based on comprehensive climate alignment. It not only enhances reporting and risk assessment but also propels value creation by unveiling competitive advantages and ensuring strategic integration within the climate economy. Crucially, UCI introduces the capability to financially quantify impacts, bridging the gap between climate data and financial materiality (FM). This financial lens is essential for decision makers, as it consolidates fragmented insights into a unified view, enabling informed decisions that drive value realization and accelerate climate action. Without such financial quantification, the journey toward meaningful climate action remains hindered, underscoring the importance of integrating UCI for holistic and effective climate strategy implementation.

However, the journey to date toward climate resilience and sustainability has been dominated by the use of fragmented single-dimensional datasets and models e.g., CO_2 accounting, physical hazard exposure and transition pathway analysis. For example, let's explore the case of CO_2-focused strategies. While CO_2 models and accounting are foundational in creating net zero strategies, they cannot by themselves address the comprehensive spectrum of climate-related risks and opportunities. Achieving net zero emissions is an essential milestone in our global efforts to combat climate change, yet it's increasingly recognized as a necessary but insufficient condition on its own. The focus on net zero, primarily targeting the reduction of CO_2 emissions to mitigate global warming, risks a myopic view that overlooks interconnected environmental, social and economic challenges posed by climate change. This includes the existing and accelerating threats of physical climate risks – like extreme weather events, sea-level rise and the heat island effect – as well as the critical issue of biodiversity loss and ecosystem degradation. Furthermore, CO_2 accounting in itself does not provide insights into the business impacts of current and future emissions under different transition scenarios in the future. Single dimensional views also lead to suboptimal financial decisions. A unified, financially quantified view is needed to determine which climate investments (e.g., adaptation, decarbonization, resource efficiency, etc.) to prioritize and in some cases to assess trade-offs among these.

The market's prevailing focus is on CO_2 accounting and mitigation, which is significantly influenced by global policy directives and national government commitments to reduce greenhouse gas (GHG) emissions. This emphasis has become a central strategy for many organizations aiming to demonstrate their commitment to climate action. Governments worldwide, from the European Union's ambitious Green Deal to China's pledge to become carbon neutral by 2060, have not only set carbon reduction targets but also established frameworks and incentives for businesses to contribute to these goals, thus shaping corporate strategies toward more sustainable practices. It's essential to recognize the significant strides made through initiatives like renewable energy adoption, energy efficiency improvements and the growth of the electric vehicle market. These actions, while focused on carbon reduction, pave the way for broader sustainability and new innovation efforts, demonstrating the tangible and powerful impact of concerted global and corporate commitment to reducing CO_2 to tackle climate change.

A key feature of these CO_2 emission reduction strategies has been carbon offsets. However, their limitations and potential for misapplication have resulted in a tumultuous journey to date. For an offset project to be truly effective, it must result in additional carbon reductions (often called "additionality") that wouldn't have occurred without the offset investment, avoid leakage where the offset project unintentionally causes increased GHG emissions somewhere outside the project boundary and have permanence so that the offsets are long lasting. However, ensuring and verifying additionality can be challenging. Furthermore, the permanence of these projects, especially those based on forestry and land use, can be uncertain due to risks such as fire, disease or changes in land management policies calling into question the durability of the offsets. The Bootleg Fire, which occurred in Oregon in 2021, serves as a poignant example of the risks associated with poor quality carbon offsets and the challenges involved in ensuring permanence of the carbon sequestration. This particular wildfire burned through many thousands of acres, including areas that had been designated for carbon offset projects highlighting the vulnerability of relying on forestry-based offsets for carbon sequestration. The outcome ultimately resulted in the destruction of 3.3 million metric tons of stored CO_2 (Bernton, 2023), negating any of the long-term climate benefits.

The Bootleg Fire illustrates the role that enhanced intelligence frameworks like UCI can play in considering the full spectrum of climate risks, not just in the immediate term but over the entire lifespan of a project or investment. A clearer understanding of physical risks, including detailed assessments of climate vulnerabilities and the likely impacts of extreme weather events like wildfires, could significantly inform the selection and management of carbon offset projects and would have led to better preparation and resilience measures, potentially mitigating the loss of carbon sequestration benefits when disaster struck.

Over-reliance on carbon offsets is counterproductive to effective climate action and risks diverting attention and resources from direct actions that reduce emissions at their source. From an organization perspective, this approach might lead to missed opportunities to achieve operational efficiencies, invest in renewable energy or innovate in product design – all of which can provide long-term value to shareholders and stakeholders alike. For instance, companies like Google, Microsoft and Apple are investing in renewable energy projects not merely to offset their carbon footprint but to power their operations sustainably, showcasing a shift toward direct action in climate strategy. It's recognized, however, that in hard-to-abate sectors, a temporary reliance on CO_2 offsets might be unavoidable until more sustainable technologies become accessible. In these contexts, "sunset clauses" for offsets – phasing them out by specified years – could ensure their use as a transitional measure rather than a perpetual fix. By setting a clear timeline for phasing out offsets, industries are encouraged to invest in and transition to sustainable technologies more rapidly. Overall, this approach emphasizes the necessity of moving beyond offsets, underscoring the importance of developing strategies that prioritize direct emission reductions, resilience-building and innovation. Such strategies better align with the goals of comprehensive climate frameworks like the Paris Agreement, which calls for a systemic reduction in global emissions and fosters resilience across ecosystems and economies.

Broadening the lens beyond single-dimension climate risk focus is just as essential when addressing physical, transition or nature risk challenges. Any approach in isolation will lead to imbalanced actions that might overlook critical vulnerabilities or opportunities carrying the risk of creating blind spots in a company's climate strategy. Examples of each of these to highlight the disconnect are:

Physical Risk: A Coastal Infrastructure Expansion

A company plans to fund and expand port facilities to increase trade capacity. In assessing project feasibility, the company primarily focuses on physical risks, particularly sea-level rise and storm surges, to ensure the infrastructure is resilient to these climate impacts. This unidimensional focus leads to the development of robust flood defences and elevated structures to mitigate the anticipated physical risks. By concentrating on physical risks alone, the company overlooks transition risks associated with shifting regulatory environments and market dynamics toward low-carbon economies. Future and probable regulations on shipping emissions or incentives for decarbonizing maritime transport could shift trade patterns or reduce the demand for traditional port services, affecting the project's long-term viability and financial returns. Socio-economic factors are also neglected, such as community displacement or changes in local employment patterns due to the expansion. This oversight can result in community opposition, reputational damage and potential delays or increased costs due to the need for additional social mitigation measures.

By not embracing a holistic approach to climate risks and opportunities, organizations overlook the chance to harness innovative solutions that span across multiple dimensions. For instance, the integration of green infrastructure can serve dual purposes: it offers flood protection while simultaneously enhancing biodiversity, which in turn benefits local communities and can even unlock new revenue opportunities, such as eco-tourism or carbon credits through initiatives like mangrove restoration. Such forward-thinking strategies not only mitigate climate risks but also contribute positively to the organization's value proposition.

Transition Risk: A Green Energy Company

The company decides to significantly invest in renewable energy projects, such as wind and solar power, divesting from fossil fuel assets to align with global climate targets and anticipate regulatory changes favouring clean energy. They channel substantial resources into developing renewable energy infrastructure, aiming to become a market leader in clean energy, as well as lobbying for favorable renewable energy policies and investing in marketing to position itself as a green energy provider.

By overlooking the physical risks posed by climate change to its new and existing infrastructure, the company fails to consider the increased risk of hurricanes and storm surges to its coastal wind farms or the vulnerability of its solar panels to extreme hail storms and temperature fluctuations. Not accounting for extreme weather conditions exposes the company to operational disruptions and an inability to guarantee 100% uptime, jeopardizing commercial contracts and accruing significant repair costs alongside revenue losses. This situation not only risks damaging stakeholder trust in the company's resilience and long-term viability but also raises serious concerns about the company's capacity to service debt obligations, particularly for projects financed through debt. The financial strain from unexpected repair costs and revenue shortfalls could significantly impact the company's debt servicing capabilities, putting additional pressure on its financial health and organization value.

Nature Risk: An Agricultural Investment Project

A project gains investment for converting conventional farms into regenerative ones, implementing practices such as cover cropping, reduced tillage and agroforestry. The project focuses on the benefits these practices have for soil health, carbon sequestration and biodiversity enhancement, anticipating that these environmental improvements will also translate into long-term profitability through healthier crops and ecosystems.

While offering many positive benefits as a nature-based solution, the project overlooks potential transition risks, such as changing market demands for organic or sustainably produced food products, or shifts in agricultural policy that could affect project viability. Serious physical climate risks such as increased frequency of droughts or floods exacerbated by climate change could also jeopardize the project's success and resilience, resulting in diminished investor confidence.

These examples clearly demonstrate the ongoing challenge that organizations face in both quantifying and planning for the multidimensional nature of climate change. In essence, it's virtually impossible for a company to fully grasp or effectively prepare for these diverse and interlinked aspects of climate risk using traditional, siloed approaches. Given the nascent state of the climate expertise talent market, acquiring comprehensive in-house expertise in this area also remains a challenge across the globe.

This is where UCI emerges as a critical and accessible solution, offering a solution that can be seamlessly integrated into existing organization decision systems, providing a more comprehensive viewpoint. This aims at not only reducing carbon footprints and mitigating the diverse array of risks, but also addresses broader environmental and social dimensions, safeguarding operations, assets and shareholder value ultimately materially improving future financial performance. However, the challenges and opportunities identified by a comprehensive assessment of climate variables are not just environmental or operational concerns but are intrinsically linked to financial performance and value creation. Recognizing the financial materiality of climate risks and opportunities serves as the cornerstone for organizations in deploying UCI.

Single-dimensional climate risk assessment models stem from a variety of scientific disciplines and as such they operate orthogonally to one another. Each provides valuable insights within its domain, yet when pieced together, they reveal only a fraction of the comprehensive risk and opportunity landscape. This fragmentation presents a significant challenge in painting a full picture of climate impacts, particularly in translating these impacts into FM. The difficulty in achieving a holistic understanding of FM, in turn, fuels uncertainty in decision making and hesitation in driving climate action. This creates a circular problem where the inability to fully grasp the financial implications of climate risks and opportunities perpetuates a cycle of strategic hesitancy and misaligned priorities.

As such and despite ample investment, the Value Translation Gap introduced in the previous chapter persists. In particular, organizational leadership including financial and sustainability officers are unable to manage their climate-related financial performance, or realize full business value due to a lack of financial quantification on climate risks and opportunity. Essentially, to accelerate climate action, decision makers need to first understand the value.

The concept of FM refers to the significance of information in the context of its potential to influence the economic decisions of users, such as investors, creditors and shareholders. In the realm of climate-related risks and opportunities, FM assesses how these non-financial factors can impact a company's financial performance, value and risk profile over the short and long term. It guides organizations in determining which climate-related issues are significant enough to warrant disclosure and consideration in their strategic planning and risk management.

The integration of FM into climate-related risk reporting and disclosures has created a critical bridge between climate strategy and financial performance. This offers a universal, comparable and standardized approach enabling organizations to move from viewing climate initiatives as cost centers to recognizing them as drivers of value creation. As such, organizations can leverage traditional financial decision making artifacts such as balance sheets, profit and loss statements, return on investment (ROI) calculations, discounted cash flows and net present value analysis to enable decisions. These tools have significant advantages to unlocking action, because they are both well-understood and scalable but also have been battle-tested across various sectors and market conditions. The deeper integration of climate analytics into business operations for strategic decision making denotes the market shift toward the new generation of organization climate alignment. This evolution emphasizes actionable insights over data collection, moving from compliance-driven to value-driven approaches and most importantly, leverages advances in technology to apply a financial lens to climate data analytics.

The first generation of climate software has been fundamental in building the foundations of the CI sector, primarily seeking to answer the "what is happening" of climate drivers by focusing on the collection and presentation of data such as flooding risk, CO_2 accounting and resource (e.g., water and energy) usage. This initial phase was largely about quantifying and reporting on sustainability metrics as well as the exposure that a company has from the direct effects of climate change, aligning with developing regulatory requirements and corporate sustainability goals. However, it falls short of integrating these metrics into the broader strategic business value or decision making processes, serving more as a compliance tool than a strategic asset. The next iteration – Climate Tech 2.0 – moves beyond data tracking and reporting, toward an integrated architecture that consolidates all climate intelligence and interprets it through a unified financial lens. This marks a significant leap from simply identifying climate risks to actively leveraging these insights for strategic advantage, guiding businesses through informed, value-driven climate actions. The shift underscores the necessity for decision makers, such as organization executives, to adopt a broader, more integrated approach to climate responsiveness, moving beyond compliance to unlock real business value. In this context, the application of frameworks like QuantEarth™, outlined in more detail below, illustrate how businesses can harness the power of Climate Tech 2.0. By providing actionable, value-centric business insights, a UCI framework can facilitate a seamless transition to this new paradigm, where climate intelligence becomes a key driver of business strategy and sustainability a source of strategic differentiation.

UCI Framework Case Study: A Comprehensive Strategy for Climate Responsiveness

The QuantEarth™ Framework (QE) was developed for practitioners by the authors (Bassi and Chopra), to reduce the barriers and complexities to aligning climate-related decisions to organization strategy. Using a wealth of expertise gained from the creation and growth of the climate tech category over the past 15 years, the aim was to create a purpose-built platform that would help practitioners to intuitively understand climate-related financial performance at the organization, business unit or product level. It is a comprehensive method for the implementation of UCI into the fabric of business operations, facilitating action across all levels and is rooted in both financial analysis and a strong understanding of policy and disclosure changes. QE uses a three-stage process to evaluate climate risk and opportunity, and deliver financially-quantified intelligence to support resilient climate action: Quantify – Plan – Act.

To deliver its unique insights, the QE Framework harnesses an extensive array of models, encompassing transition pathways, physical hazard assessments, natural capital indicators and energy consumption metrics.

This integration leverages peer-reviewed scientific models and datasets, including CMIP6, CORDEX, Aqueduct, NASA's Earth Exchange Global Daily Downscale Project (NEX-GDDP), CaMa-Flood and others ensuring a foundation in cutting-edge climate science. Moreover, the framework aligns with prevailing policy frameworks at global, national and sector-specific levels. It incorporates transition models, such as the IEA's Global Energy and Climate Model, NGFS Integrated Assessment Models as well as specialist datasets spanning climate policy, legislation, water pricing, natural capital and more. These foundational models, along with scientifically rigorous modeling and financial analytics, translate climate drivers – across physical, transition and natural capital for multiple timescale and scenarios – into financial impact. These include well-recognized accounting tools for strategic decision making, such as Profit & Loss statements, Discounted Cash Flow analysis and Earnings Per Share calculations, facilitating a comprehensive and actionable financial perspective on climate-related business performance. Corporate organization data, provided by the user, including key business units, financial statements, geo-distributions of revenues, supply chain dependencies, sales projections, emissions data and asset location, is also integrated. QE then uses an array of analytical tools and ML approaches to fuse this organization data with its climate models to provide analyses and probabilities across scenarios and business relevant timeframes. This fusion allows users to see their financial materiality from a "whole of the organization view" right down to a single asset like a warehouse or a factory in its supply chain.

Stage 1 – Quantify

QE covers the three main principles of (i) a unified, multidimensional approach with a 360° view of climate-related factors; (ii) the translation of each climate factor into financially quantified metrics for different timescales and climate scenarios; and (iii) tailored organization analysis down to a granular view, retaining fidelity at every level of interrogation.

By taking this holistic approach, the initial assessment looks at the financial impact of climate drivers on the organization. This encompasses a vast range of variables, including the financial repercussions of direct damage, business interruptions and revenue implications, as well as productivity loss and increased operational costs resulting from extreme weather. It also accounts for the economic effects of transitioning toward a low-carbon economy, which covers regulatory costs, shifts in consumer preferences, capital availability, technological adoption and variations in input prices. An organization must be enabled to gain insight into the interconnections between potential actions for effective decision making – reducing the impact of heat stress on productivity may involve, for example, increasing the cost of energy and carbon taxes paid by implementing temperature control. By offering scenario analysis, the framework allows organizations to explore various future landscapes under different climate scenarios, both physical and transition. This step moves beyond simply identifying "what is happening" to understanding the "so what" – clarifying the potential drivers of future climate-related financial performance, their magnitude, timing and conditions. Crucially, the process simultaneously identifies opportunity, recognizing the potential for revenue enhancement, cost reduction, asset valuation increases and resilience building through climate-aligned actions and market opportunities. It is essential that value creation remains at the core of UCI, mirroring the long-term investment that must be made into climate resilience.

Such financially quantified insights travel far beyond what has been traditionally available to decision makers. For the first time, UCI frameworks such as QE can assess contribution and materiality, understanding which business unit, product line, geographic region or supplier carries the most risk, or which climate drivers have the biggest impact on value creation over different time scales. Combined with the race toward radical transparency, benchmarking future financial performance against competitors or a peer set becomes quantifiable and essential to plan strategic competitiveness. Fundamentally, without quantification, planning and action cannot be executed empirically.

Stage 2 – Plan

Building upon a robust understanding of climate-related financial impacts in the "Quantify" stage, the "Plan" stage strategically navigates through adaptation and efficiency enhancements, defining clear ROI criteria. This stage empowers organizations to weave climate considerations into their financial forecasting and strategic planning processes, including projections of future earnings and organization value. It's about turning insight into action, determining the most strategic paths for climate resilience and value creation across the organization – the "then what," which follows the "so what" of quantification.

Crucially, the Plan stage enables companies to evaluate trade-offs and optimize decisions with a strategic lens. For instance, given a finite budget or capital pool, organizations can use UCI to identify the most impactful climate actions – be it in resource efficiency improvements, adaptation measures, or direct emissions mitigation strategies. This process involves assessing the ROI across various climate actions, helping businesses prioritize initiatives based on their potential to enhance resilience, reduce costs or generate new revenue streams. The granularity of UCI analysis facilitates bespoke decision making across different operational domains, providing the detailed insights necessary for decision makers to align their strategies with both corporate climate goals and local operational realities. This stage dispels the notion of a "one size fits all" approach, acknowledging the unique challenges and opportunities faced by different areas of an organization.

Stage 3 – Act

With a comprehensive understanding of climate impacts achieved through the initial stages of quantify and plan, organizations can now move into the critical phase of action. The "Act" stage is where UCI analysis transforms into tangible, actionable strategies to deliver value such as bolstering resilience and enhancing operational efficiency. This phase is about turning insights into impact, ensuring that the myriad climate-related variables identified are met with equally multifaceted solutions.

The actionable strategies emerging from UCI encompass a broad spectrum, addressing both the immediate and long-term needs of an organization. From infrastructure enhancements designed to withstand extreme weather events to shifts in supply chain management aimed at reducing carbon footprints, the solutions are as varied as the challenges they aim to solve. Importantly, these actions consider both the direct costs and the broader implications across the value chain, ensuring that decisions are made with a holistic view of their impact.

The "Act" stage of the QuantEarth™ Framework is significantly enhanced by its provision of a segmented view of opportunities and risks. This level of detail allows for the efficient and targeted prioritization of resources – for instance, identifying that adaptation may be required at Factory Z, while a nature-based intervention might be necessary within a specific segment of the company's supply line. Such granular insights, coupled with robust attribution analysis, facilitate the internal construction of compelling business cases for investment.

Integrating resilient climate action across the organization necessitates a rigorous quantification of risks and opportunities, alongside the formulation of a clear, organization-wide climate prioritization plan. The granularity of the insights from UCI empowers decision makers at all levels to take informed actions. This includes providing aggregated insights for C-level executives and the board to set company-level targets and investments, as well as offering detailed, product or regional insights for individual business unit owners to enhance their specific operations. By adhering to the structured framework of quantify, plan and act, businesses transition from fragmented or reactive approaches to a strategic pathway that fosters authentic climate value creation. This method ensures investments in climate resilience and adaptation are both strategic and aligned with the organization's overarching goals, promoting sustainability and value creation that echoes throughout the entire value chain.

Finally, QE is designed to support the regulatory and stakeholder demands for transparency in climate risks, opportunities and action. It has become crucial for organizations to articulate their climate journey. The new wave of disclosures, such as CSRD and standards set by the IFRS, mandate a more penetrating analysis. These regulations emphasize the importance of financial materiality and the identification of opportunities for value creation. Compliance with these standards necessitates a "model fusion" that combines insights from climate sciences, transition pathways and their integration into financial decision making tools. Without this comprehensive approach, meeting the demands of these disclosures and demonstrating a commitment to sustainability becomes a challenge. Reporting should transcend compliance, serving as a platform to showcase the strides made toward meaningful change. By documenting and sharing the progress made from quantification to action, organizations not only meet disclosure requirements but also demonstrate their commitment to sustainability and long-term value creation. In this way, climate action is repositioned from a mere regulatory obligation to a cornerstone of strategic advantage, ensuring that companies are transformed from passive climate responders into proactive architects of climate-resilient value.

Importantly, Gen-AI enables anyone within the organization to interrogate and engage with the QE insights in natural language as well as automate workflows and easily integrate insights into their workflows and decisions. The Climate Gen-AI – powered by QE – is an expert assistant making specialist insights accessible without the need to be a climate scientist, data engineer or financial analyst. Anyone in the company can now learn, understand, analyze and leverage UCI. This democratizes UCI and gives everyone the power to be able to apply it in their role and ultimately make better decisions that drive value and accelerate climate action within companies.

Reference

Bernton, H. (2023, 2 August). A giant Oregon wildfire shows the limits of carbon offsets in fighting climate change. OPB. https://www.opb.org/article/2023/08/02/climate-change-carbon-offset-oregon/

CHAPTER 9
UCI FOR COMPETITIVE AND COLLABORATIVE ADVANTAGE

Summary

We explore the role of Unified Climate Intelligence (UCI) in redefining competitive advantage for organizations operating within the burgeoning climate economy. UCI empowers businesses to navigate the complex interplay of market forces and environmental challenges, enabling them to identify and seize opportunities for sustainable growth and resilience. By integrating UCI with established strategic frameworks like Porter's Five Forces, businesses can gain deeper insights into competitive dynamics, factoring in not only traditional business forces but also physical, transitional and nature related risks and opportunities. This integration allows companies to forge robust competitive strategies that are informed by comprehensive climate intelligence.

We highlight the critical role of UCI in enhancing scenario planning and strategic foresight, exploring how UCI equips businesses with the tools to anticipate and adapt to various future climate scenarios, ensuring growth, resilience and sustainability in their operations. Furthermore, we cover how UCI redefines business-to-business (B2B) collaboration, leading to more innovative and sustainable business practices. Through UCI-driven insights, companies can build more resilient supply chains, drive efficiencies and foster B2B collaborations that are not only environmentally sustainable but also yield significant financial benefits. As we continue to navigate the challenges posed by climate change, UCI becomes an indispensable asset for businesses seeking to thrive in a competitive, climate-aware market.

As climate change increasingly influences market dynamics and shapes consumer expectations, the significance of the climate economy for corporate strategy and market positioning has never been more pronounced. Today's organizations find themselves navigating several environmental challenges, evolving regulatory frameworks and a growing demand for sustainable practices. In this evolving landscape, Unified Climate Intelligence (UCI) emerges as a critical tool for businesses committed to resilient growth. Defined as an all-encompassing approach to evaluating and acting upon climate-related risks and opportunities, UCI equips businesses with the insights needed to navigate the complexities of physical, transitional and natural risks and resource usage, alongside the broader socio-economic implications of climate change.

Adopting UCI enables companies to identify hidden opportunities for operational efficiencies, drive innovation and enhance resilience against climate-induced vulnerabilities. It provides a strategic foundation for anticipating regulatory shifts, aligning with emerging consumer preferences and pre-emptively managing physical climate risks with a high degree of confidence. This chapter aims to articulate how UCI can be leveraged as a competitive driver, offering businesses the foresight and agility needed to excel within the dynamic climate economy.

Competitiveness, a critical determinant of an organization's success and longevity, has always propelled businesses to seek innovative strategies and frameworks to outperform rivals. Throughout the late twentieth century, traditional frameworks have been pivotal for organizations aiming to decipher and navigate the complexities of their competitive environments. These models, encompassing a variety of analytical tools and methodologies, have guided businesses in dissecting market dynamics, assessing competitive threats and identifying opportunities for strategic positioning and value creation. As the foundation of strategic planning, these frameworks enabled a structured approach to understanding market forces, customer behavior and the broader economic landscape. One of the most popular and highly regarded business strategy tools to date is Porter's Five Forces, which provides a strong foundation for understanding competitive dynamics. It breaks down the forces shaping competition into five distinct categories: the threat of new entrants, the bargaining power of suppliers, the bargaining power of buyers, the threat of substitute products or services and the intensity of competitive rivalry. Each force plays a critical role in determining the strategic direction and profitability of businesses within an industry.

Moving into the twenty-first century, the rapid pace of technological innovation, globalization and evolving consumer expectations are continually shaping the business environment. These shifts have brought new challenges and opportunities to the forefront, highlighting the need for traditional strategic models to adapt and incorporate the realities of digital transformation, sustainability and social responsibility. By adjusting these frameworks to reflect the latest market trends and environmental considerations, businesses can stay competitive, foster innovation and build resilience against future uncertainties. The integration of UCI with conventional strategic frameworks introduces a refined perspective on competitive dynamics, enriched by an awareness of climate implications. The shift from a focus solely on competition to a broader understanding of the changing global landscape enables companies to navigate toward sustainable growth and long-term value creation more effectively.

By integrating UCI into a framework such as Porter's Five Forces businesses can anticipate and adapt to emerging trends and vulnerabilities, positioning them to navigate the climate economy with greater agility and foresight. The following examples illustrate how UCI's comprehensive insights enhance understanding and strategic planning within these areas.

Competitive Rivalry: examines the intensity of competition among existing firms in the industry. High competitive rivalry can limit profits and push companies to innovate to maintain or gain market share.

Without UCI: A company might focus on traditional competitive strategies such as cost reduction and product differentiation without considering sustainability factors, potentially missing out on emerging customer preferences for eco-friendly products. Furthermore, a company may fail to anticipate impacts on margins of market shifts (e.g., changes in the price of renewable versus traditional energy sources due to policy changes and technological innovation), eroding margins and competitiveness in the future.

With UCI: Leveraging UCI, the company conducts a thorough benchmarking of its climate performance against competitors, discovering opportunities to separate from the competition. It uses this insight to market itself as a sustainability leader, attracting eco-conscious customers and investors and distinguishing the company in a saturated market. With UCI, a company can also anticipate the impact on margins of future policy and technology scenarios (e.g., changes in price of different energy sources, future carbon taxes and falling price of low-carbon technologies) to innovate and enhance margins, and thus its competitive position.

Bargaining Power of Suppliers and Buyers: considers the impact of suppliers and buyers on business practices. The bargaining power of suppliers determines how much control they have over the price and quality of goods and services, potentially squeezing profitability for companies. Conversely, the bargaining power of buyers assesses the influence customers have on price and product offerings.

Without UCI: A company might not fully assess the risks associated with its suppliers, such as water scarcity impacting raw material availability, leading to potential supply chain disruptions and increased costs. Additionally, without understanding the evolving policy landscape on carbon pricing and taxes or supply chain emissions requirements of its largest customers, a company can end up being uncompetitive compared to other potential suppliers.

With UCI: Integrating UCI, the company evaluates climate vulnerabilities, like water scarcity or susceptibility to extreme weather events, within its supply chain, identifying and engaging with suppliers in lower-risk areas or those employing sustainable practices. This proactive strategy not only secures the supply chain but also potentially negotiates better terms. Integrating UCI also enables the company to manage risks and opportunities from the evolving policy and market landscape around carbon prices, ensuring that it remains competitive, meets buyer requirements and/or commands a premium through differentiation. This approach strengthens the company's market position by ensuring supply chain resilience and aligning with dynamic market and policy changes.

Threat of Substitution: evaluates the likelihood of customers switching to alternative products or services, posing a threat to a company's profitability. It prompts businesses to assess the availability and attractiveness of substitutes in the market.

Without UCI: A company may not fully recognize emerging alternatives that are less climate impacting, thereby missing opportunities to adapt its offerings or invest in innovation to stay competitive. This oversight could result in declining market share as consumers shift toward more sustainable options provided by competitors.

With UCI: Leveraging UCI, the company identifies emerging substitutes with lower climate impact, allowing it to adapt its offerings or invest in innovation proactively. By staying ahead of market trends and aligning with consumer preferences for eco-friendly products and services, the company maintains its relevance and competitive edge. Furthermore, by investing in sustainable innovation, such as renewable energy sources or eco-friendly materials, the company not only mitigates the risk of substitution but also enhances its brand reputation and attractiveness to environmentally conscious consumers.

Threat of New Entry: assesses the ease with which new competitors can enter the market, potentially intensifying competition and reducing profitability for existing players. It considers factors such as barriers to entry, economies of scale and brand loyalty.

Without UCI: Without considering climate-related factors, a company may overlook significant barriers that new entrants face, such as stringent environmental regulations or the need for sustainable infrastructure. Consequently, the company may underestimate potential competitive threats and fail to leverage its existing advantages in climate resilience or sustainable practices.

With UCI: By integrating UCI into strategic analysis, the company identifies climate-related barriers that new entrants may encounter, such as the need for sustainable infrastructure or compliance with stringent environmental regulations. Moreover, UCI helps to identify climate-related risks that new entrants might overlook, such as flood-prone areas for potential production facilities. Recognizing these barriers and risks allows the company to leverage its existing investments in climate resilience as a competitive advantage, making it more challenging for new entrants to establish themselves in the market. Additionally, the company can proactively invest in sustainable practices and infrastructure, further strengthening its position and deterring potential competitors.

Building business models for the climate economy is now a strategic necessity. These models should integrate climate considerations into core business operations leading to resilience value creation. Alongside powerful insights driven by UCI, is the adoption of circular economy principles, which aim to redesign the traditional "take-make-waste" model into a more regenerative and restorative system, working to detach economic activity from the consumption of finite resources, and designing waste out of the system.

At the heart of the circular economy are three guiding principles: designing out waste and pollution, keeping products and materials in use and regenerating natural systems. The first principle involves rethinking how resources are used – products are designed from the outset to use materials that are safer and more sustainable, and that can either biodegrade naturally and safely or be repurposed and recycled indefinitely. The second principle focuses on maximizing the usability of products. This is achieved by designing for durability, repairability and recyclability. Businesses may shift from selling products to leasing them, thereby retaining ownership of the materials used and ensuring that they are handled responsibly at the end of their life. Third is adopting practices that restore and rejuvenate the environment. This includes using production methods that replenish and repair natural ecosystems rather than deplete them. By doing so, businesses can operate within the planet's ecological limits, which as we covered in Chapter 2 are far exceeded globally at the present time.

Around the world, companies are increasingly recognizing the benefits of sustainable business models. Innovations such as bio-based materials, digital technology for resource efficiency and blockchain for enhancing transparency are becoming more prevalent. These trends are supported by a growing body of research that demonstrates the long-term profitability and resilience of businesses that commit to sustainable practices. Investing in sustainable practices offers multiple financial benefits. By optimizing resource use and reducing energy consumption, companies can achieve substantial cost savings. For instance, transitioning to energy-efficient systems or adopting renewable energy sources like solar and wind power significantly reduces pricing volatility and utility expenses. Moreover, these sustainability efforts often qualify businesses for green financing – preferential loans, grants and other financial incentives designed specifically to support environmentally positive projects. Such financial products are typically more advantageous than conventional options and can make a considerable difference in the feasibility and impact of sustainability initiatives.

Companies that embrace sustainable practices tend to experience an enhancement in brand reputation, which can lead to increased customer loyalty and even premium pricing opportunities. Today's consumers are increasingly aware of and concerned about environmental issues and often prefer to support businesses that demonstrate a commitment to sustainability. Furthermore, proactive adoption of sustainability measures positions companies favourably concerning regulatory compliance, reducing the risk of fines and penalties associated with environmental breaches.

As the global business environment races to adapt to the climate challenge, the ability to foresee and prepare for various future scenarios is essential for maintaining resilience and competitiveness. UCI is instrumental in this regard, equipping businesses with the tools needed for dynamic scenario planning. This approach leverages detailed climate data and predictive analytics to envision potential future states and their impacts on operations, enabling companies to develop adaptive and forward-looking strategies. In this case, UCI is particularly valuable for businesses considering a transition to circular models as it can provide insights into how these models might perform under different environmental, economic and regulatory scenarios.

UCI-enabled scenario planning allows organizations to integrate the evolving policy and market landscape around carbon pricing into their decision making. For example, by looking at the carbon prices under potential future scenarios (e.g., for the world to get to net zero by 2050), a company can better understand the impacts on margins of its current emissions and the value of decarbonization. It can also understand the differences across its operating regions and prioritize investments accordingly. Similarly, a company can project future changes in energy prices of renewables versus traditional sources across its different business units and operating regions to estimate future cost and margin impacts.

UCI prepares businesses for climate-related contingencies. As extreme weather events and resource scarcities become more frequent, strategies that incorporate these risks are indispensable. UCI can enable organizations to simulate the effects of various climate scenarios on their supply chains and production processes, identifying vulnerabilities and developing mitigation strategies such as alternative materials or suppliers before they become urgent and expensive. By incorporating UCI into their strategic planning processes, businesses can enhance their ability to navigate a future marked by uncertainty. This capability not only secures a company's operational resilience but also provides a competitive edge in a market where agility and foresight are increasingly crucial.

The shift toward sustainability is not only reshaping consumer markets but also transforming business-to-business (B2B) relationships. Here, UCI plays a crucial role in this transformation by enabling companies to integrate circular economy principles into their B2B strategies. These principles focus on extending product life cycles, enhancing material reuse and fostering collaborations that are both economically and environmentally sustainable.

UCI aids companies in identifying and managing the risks and opportunities across the value chain. For instance, through UCI, businesses can gain insights into hidden costs and opportunities upstream and downstream. This intelligence is crucial for forming strategic alliances that can leverage shared resources, technology and expertise to enhance overall sustainability and drive business value. This alignment is essential for improving and maintaining brand integrity, meeting consumer expectations for sustainability and creating resilient supply chains that are less susceptible to disruptions from resource scarcities or regulatory changes.

UCI facilitates the sharing of materials and resources among businesses, which is a fundamental aspect of the circular economy. By using UCI to track material flows and lifecycle data, companies can create closed-loop systems where waste from one process becomes the input for another. This not only minimizes waste and reduces environmental impact but also leads to cost reductions by optimizing resource use. BMW and Jaguar Land Rover have embraced these principles by partnering with recycling firms to reuse materials from end-of-life vehicles. This practice not only conserves resources but also reduces production costs and minimizes the environmental footprint of new cars. Incorporating UCI into B2B relationships enables companies to create networks that are not just supply chains but value circles – where every participant benefits from shared commitments to efficiency and sustainability. These partnerships are characterized by a transparency that allows all parties to monitor progress toward environmental goals, fostering a collective responsibility toward global sustainability.

Strong supply chains are a crucial component of business resilience. Businesses face a multitude of risks, from natural disasters and geopolitical tensions to more frequent and severe climate events. Here, UCI can enhance supply chain resilience by identifying potential disruptions and supporting organizations to adapt for these challenges through strategic planning. For example, UCI can enable companies to conduct detailed risk and opportunity assessments, considering various climate scenarios and their potential impacts on supply chain operations. This proactive approach allows businesses to develop contingency plans, such as diversifying their supplier base or investing in more robust infrastructure to withstand adverse conditions. Additionally, UCI can identify trends and shifts in global trade patterns (e.g., CBAM), helping companies to anticipate changes in the regulatory environment or in the availability of resources.

Integrating circular economy principles into the supply chain offers another layer of resilience. Local sourcing reduces dependency on distant suppliers and minimizes logistics-related vulnerabilities and costs. Modular design, where products are designed to be easily disassembled and reassembled, allows for quicker adaptation to changes in consumer demand or supply chain disruptions. These practices not only help in managing resources more efficiently but also in reducing environmental impact, thus aligning with broader sustainability goals. Unilever is a strong example of an international organization that has actively implemented circular economy principles across its vast network of supply chains to enhance resilience and sustainability. One of the company's key initiatives is the Sustainable Agriculture Code, which aims to source raw materials in a way that minimizes environmental impact while supporting local economies. This approach includes partnering with local farmers and suppliers to reduce the distance food travels from farm to consumer, thereby decreasing transportation emissions and costs. This local sourcing strategy not only helps mitigate risks associated with global supply chain disruptions – such as those caused by extreme weather events or transportation bottlenecks – but also ensures a fresher, more reliable product supply. By investing in sustainable farming practices, Unilever aids in preserving the local environment and biodiversity, which contributes to the long-term stability of food supplies. In addition, Unilever employs modular approaches in its product formulations and packaging, enabling the company to adapt quickly to changes in consumer demand or resource availability. This flexibility is vital for maintaining supply chain continuity and for responding dynamically to market or environmental shifts both locally and globally.

As organizations increasingly recognize the interconnectedness of their operations with broader environmental and societal impacts, B2B collaborations are evolving. UCI and the principles of the circular economy are playing pivotal roles in shaping these new business relationships. These tools and concepts help companies align their strategies not just for mutual profitability but also for sustainability and resilience. This capability is crucial for developing partnerships that go beyond traditional supply chain transactions and into more collaborative, innovative and sustainable practices. Companies could, for example, use UCI to share data on resource usage and waste generation, identifying opportunities to reduce, reuse or recycle materials jointly. As discussed in earlier chapters, the shift toward radical transparency is becoming an inevitable part of doing business, driven by advances in technology and heightened global awareness. This transparency not only ensures that all stakeholders can see and understand the impacts of business activities on the environment and society but also drives competitive behavior as companies can see each other's data, fostering a race to the top in sustainability practices.

In the context of circular economy principles, the future of B2B collaboration could see businesses sharing facilities, logistics and even crucial information systems to optimize resource use and minimize waste. This shared use of infrastructure not only reduces costs but also lessens the environmental impact associated with manufacturing and distributing products. The technology sector is already embracing this opportunity, with companies like HP and Dell participating in material consortiums to share rare materials reclaimed from recycled electronics. This approach not only ensures a more stable supply of scarce resources but also reduces the environmental impact associated with mining new materials. In addition, these collaborations extend to sharing best practices for product design that facilitates recycling and end-of-life processing, ensuring that products are designed with their entire lifecycle in mind. By doing so, companies can create a closed-loop system where materials are perpetually recirculated within the industry, reducing the need for raw materials and decreasing the overall environmental footprint.

In the future, the potential for B2B collaborations to foster sustainable growth is immense. Companies that embrace these opportunities can lead the way in creating an economy that is not only more efficient and competitive but also more attuned to the planet's ecological limits and societal needs. This shift is not just beneficial in terms of corporate sustainability goals but is also increasingly becoming a competitive differentiator in industries worldwide.

We will conclude this chapter with a series of case studies that illustrate the practical application and benefits of UCI across different industries. These real-world and illustrative examples from the mining, fashion and automotive sectors demonstrate how integrating climate intelligence into business strategies can enhance competitive positioning, foster innovation and drive sustainable growth.

Illustrative Application of UCI to Mining Sector: *BHP*

BHP, formerly known as BHP Billiton, is a leading global resources company headquartered in Melbourne, Australia. It ranks among the world's largest mining companies, focusing on the exploration, development, and production of a diverse range of commodities, including iron ore, copper, nickel and coal. BHP is publicly traded, with listings on major exchanges in Australia (ASX), London (LSE) and New York (NYSE). In the fiscal year 2023, BHP reported revenues of approximately $53.8 billion with a profit margin of about 25%. This financial health reflects the company's effective strategic management and its ability to capitalize on the robust demand for essential minerals and resources.

BHP faces a complex web of climate-related business drivers that profoundly impact its operations and strategic decision making, including an array of financial risks and opportunities shaped by the volatile forces of climate change, regulatory demands and market transformations.

Physical Risks, including severe disruptions from cyclones such as Cyclone Veronica in 2019, have damaged infrastructure and impacted production in Western Australia. Flooding, like the 2011 incidents in Queensland, has led to temporary shutdowns, affecting coal exports significantly. Additionally, the operations endure extreme heat stress, particularly in the Pilbara region, necessitating modifications in work schedules and cooling systems to maintain safety and productivity.

Regulatory Changes also pose significant challenges, with emission regulations and carbon pricing impacting operational costs. The introduction and adjustments in Australia's Carbon Pricing Mechanism have compelled BHP to invest in carbon reduction and renewable energy projects. This regulatory environment also offers opportunities to enhance competitiveness through strategic investments in green technologies.

Market and Technological Shifts are profoundly reshaping demand within the sector. The energy transition is decreasing demand for products like thermal coal, while simultaneously boosting the need for critical minerals such as nickel and lithium, essential for the burgeoning electric vehicle market.

Nature and Biodiversity issues, particularly water scarcity in South American operations, are pressing. The Escondida mine's reliance on dwindling Andean glacier water has driven investments in desalination and water recycling to secure water supplies. The intensive demand for water for mining operations not only strains the local water supply but also heightens risks related to habitat loss and community livelihoods.

Amid these challenges, BHP is working to identify opportunities to drive resource efficiency, particularly in energy and water use, and to command a green premium with value chain partners. Innovative partnerships and technologies, such as those supporting the decarbonization of the steelmaking sector, are crucial for leveraging BHP's production of iron ore and aligning with global sustainability goals.

Here we use the UCI analytics engine – QuantEarth™[1] – to quantify the financial and business impacts on BHP of the various drivers. The insights, generated using publicly sourced data, highlight the future impacts of inaction and assess the future benefits of actions BHP can take.

The financial Cost of Inaction: modeling a subset of the climate drivers – namely future carbon pricing from direct emissions (Scope 1 and 2) and energy cost based on current mix of energy sources under different transition scenarios as well as business disruption impacts from chronic physical risks such as increasing heat stress – shows up to $7.8 billion of incremental cost by 2035 if no mitigation or adaptation actions are taken. This is an increase in costs of more than 14% of FY 2023 revenue, and this is only from only a subset of the climate drivers outlined above. Even with significant revenue growth over the next decade, that kind of incremental cost will materially erode margins and equity value. This quantifies the costs of doing nothing and highlights the value in accelerating mitigation and adaptation investments.

The Business Value of Early Action & Investment: the estimated price of carbon (averaged across seven different NGFS transition scenarios) grows 215% and 146% in Australia and Chile respectively between 2030 and 2035. Thus, investments that reduce emissions faster not only have outsized planetary impact but also yield significantly greater business value.

Scenario Analysis for Future Proofing: UCI enables scenario analysis and scenario planning for multiple possible futures. For example, in a scenario where current policies stay as is or the policy response is too little, too late, the costs associated with the transition drivers will be less than half the costs of a policy scenario that is aligned to a net zero by 2050 world. However, a current policy scenario will also leave the company subject to higher costs from physical climate volatility. UCI enables you to analyze impacts, assess trade-offs and develop plans for a range of possible future scenarios.

Strategic and Tailored Decision Making with Clear ROI: UCI provides highly granular and tailored insights at the business unit, asset and geo level, enabling granular and tailored decision making. For instance, the climate drivers impacting their Australian iron ore operations are different than the ones impacting the financial performance of their Chilean copper mining operation. Highly contextual, location and asset level intelligence enable more optimal decision making. For example, heat stress is a chronic risk and often a hidden cost as it is not quantified, yet has significant impact on operations, including on worker productivity, worker safety, critical machine failure, increased energy consumption and more. While heat stress does not have the largest financial risk for BHP overall, it is one of the most material risks to Western Australia Iron Ore (WAIO) business in the short to medium term. Using scientific studies that estimate the impact of heat stress, WAIO will incur an estimated average cost of $550M per year in the short-term due to heat and humidity extremes. This granular insight at the operational level enables the business unit owner at WAIO to prioritize the right initiatives (in this case adaptive capacity to deal with heat waves) for that business unit alongside the overall company priorities around decarbonization. Adaptation and mitigation have to go hand in hand, and UCI provides the insights to prioritize which actions to take, in what timeframe and for which business unit or product line or region to maximize impact. It also enables leaders across the organization to quantify the ROI of actions they can take and make the business case for accelerated climate action that will drive resilience, value and competitiveness. These insights and recommendations are illustrated in Figure 9.1, an example output from QuantEarth™.

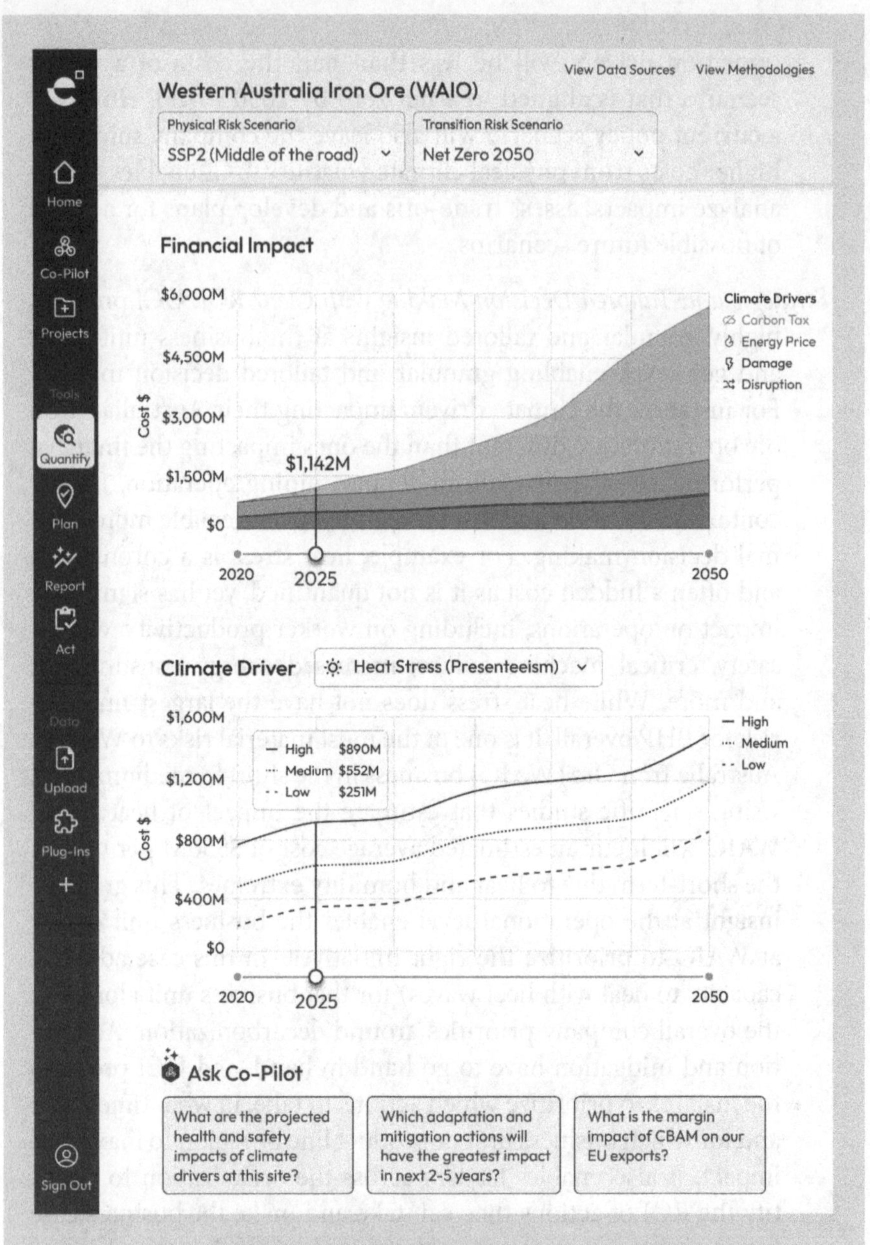

Figure 9.1 Example output from QuantEarth™.

Co-Pilot

As the site manager at Western Australia Iron Ore, how can I reduce heat stress effects in the short-term (1-2 years) and medium-term (2-5 years)? Please outline options, costs (% of revenue), and payback timelines.

Short-term (Next 1-2 years)

For immediate heat stress mitigation at Western Australia Iron Ore (WAIO), consider low-cost and quick-to-implement solutions. These include hydration stations, heat stress training, and personal protective equipment (PPE).

1. Personal Protective Equipment (PPE)

- Cost: 0.5-1% of revenue
- Break-even: Within one year
- Description: Equip workers with lightweight, heat-resistant PPE like cooling vests and breathable fabrics. This helps in temperature regulation and reduces the risk of heat stress.

2. Hydration Stations and Cooling Areas

- Cost: 0.1-0.3% of revenue
- Break-even: Immediate
- Description: Set up hydration stations and shaded cooling areas. This is a quick solution to ensure workers stay hydrated and can cool down, reducing heat-related illnesses and maintaining productivity.

3. Adjusted Scheduled and Heat Stress Training Programs

- Cost: <0.1% of revenue
- Break-even: Immediate to short-term
- Description: Educate workers on recognizing heat stress signs, proper hydration, and scheduling work during cooler hours. This training can prevent heat-related incidents and ensure quick action when needed.

Medium-term (2-5 years)

For sustainable heat stress reduction, focus on more substantial investments such as smart cooling systems, automation, and green infrastructure.

1. Smart Cooling Systems

- Cost: 1-3% of revenue
- Break-even: 2-3 years
- Description: Install misting fans and provide air-conditioned break rooms. Placing these strategically in high-traffic and high-worklo

Ask your Climate AI Expert

Figure 9.1 (Continued)

Leveraging UCI as decision intelligence at BHP allows for the democratization of climate insights, empowering employees across all roles. Foundational large language models, enhanced with specific scientific, financial and organization-specific data, allow every employee to tap into expert-level climate intelligence relevant to their roles.

CFO: Analyze the financial impacts of climate on earnings over the next three years, assess risks related to asset stranding and evaluate the consequences of delayed decarbonization in operations.

CSO: Identify the most financially significant climate risks and opportunities by region and business unit, quantify the business case for sustainability initiatives and strategize capital allocation for maximum impact under different policy scenarios.

Business Unit Owner: Explore how climate regulations and extreme events might affect production, identify supply chain vulnerabilities and evaluate the premium pricing potential of sustainable mining practices in nickel operations.

Site Manager: Determine the ROI for infrastructure upgrades to withstand severe weather, explore strategies to maintain worker safety during heat waves and optimize the allocation of capital expenditure across adaptation and mitigation efforts.

By leveraging UCI, BHP can gain precise financial insights into how climate-related factors impact its business operations, supply chains and market standing, which helps in identifying both risks and opportunities. This clarity supports strategic decision making, allowing the company to enhance revenue, minimize costs, adapt and strengthen its competitive edge. Furthermore, UCI aids BHP in proactively adhering to regulatory demands and fulfilling climate disclosure requirements, which boosts confidence among investors and stakeholders. This proactive approach ensures that BHP is well-positioned to thrive in the new climate economy, showcasing its readiness and resilience in facing future challenges.

Illustrative Application of UCI to Automotive Sector: *Electra Motors*

The automotive industry is undergoing rapid transformation due to electrification, autonomous vehicles and stricter emissions regulations. Climate change is exacerbating supply chain disruptions, resource scarcity and extreme weather events, impacting production, logistics, consumer behavior and regulatory landscapes. Heavy investments in R&D, battery technology and charging infrastructure, coupled with potential liabilities from climate-related incidents and regulatory fines, are squeezing profit margins. Additionally, changing consumer preferences and intensifying competition are impacting market share and revenue growth. Electra Motors can leverage UCI to:

Optimize Global Supply Chain: Identify optimal locations for battery production facilities based on a comprehensive analysis of energy costs, resource availability, climate resilience, geopolitical risks, transportation logistics and labor costs. Conduct scenario analysis to assess the impact of potential disruptions (e.g., natural disasters, trade wars, material shortages) on supply chain operations. For example, by modeling the impact of lithium-ion battery prices and availability on production costs, Electra Motors can optimize sourcing strategies and invest in alternative battery technologies.

Mitigate Climate-Related Risks: Quantify the financial impact of extreme weather events (e.g., hurricanes, floods, heat waves) on manufacturing facilities, distribution centres, and supply chain operations. Develop risk mitigation strategies, including insurance, relocation and supply chain diversification. For instance, by assessing the vulnerability of coastal manufacturing plants to sea-level rise, Electra Motors can invest in flood protection measures or explore alternative production locations.

Accelerate EV Adoption: Analyze consumer preferences, charging infrastructure needs, and government policies to optimize product development and marketing strategies. Identify high-potential market segments for electric vehicles based on factors such as population density, income levels, environmental awareness, and charging infrastructure availability. For example, by modeling the impact of different charging infrastructure scenarios on EV adoption rates, Electra Motors can optimize its investment in charging stations.

Financial Modeling and ROI Analysis: Conduct comprehensive cost-benefit analysis of electrification strategies, including battery technology, charging infrastructure, supply chain optimization, and marketing investments. Evaluate the financial impact of these initiatives on profitability, cash flow, return on investment and shareholder value. For example, by modeling the impact of different battery chemistries on vehicle performance, cost and environmental impact, Electra Motors can make informed decisions about R&D priorities and supplier partnerships.

Integration with Business Processes: UCI can be integrated into Electra Motors' existing supply chain management system, organization enterprise resource planning (ERP) software and financial planning and analysis (FP&A) tools. For example, UCI can feed data into Salesforce for sales and marketing teams to identify high-potential EV markets, or integrate with procurement systems to optimize supplier selection based on sustainability criteria.

By leveraging UCI, Electra Motors can gain a competitive advantage by making data-driven decisions, mitigating risks and capitalizing on emerging opportunities in the electric vehicle market.

Illustrative Application of UCI to Fashion Sector: *Fashion Flair*

The fashion industry is facing increasing pressure to reduce its environmental impact, including water consumption, waste generation and carbon emissions. Consumer awareness of sustainability is growing, leading to a demand for ethical and eco-friendly products. At the same time, global supply chain disruptions and fluctuating material costs pose significant challenges to profitability. Rising material costs, regulatory fines and reputational risks associated with unsustainable practices are impacting the fashion industry's bottom line. Additionally, changing consumer preferences and the need for investments in sustainable technologies are putting pressure on profit margins. Fashion Flair can leverage UCI to:

Optimize Supply Chain Sustainability: Identify sustainable material sources, assess supplier environmental performance and optimize production processes to reduce water consumption, energy usage and waste generation. By analyzing the financial impact of environmental metrics related to different materials and production methods, Fashion Flair can make informed decisions about product sourcing and development.

Mitigate Climate-Related Risks: Assess the impact of climate change on supply chain operations, including factors such as extreme weather events, natural disasters and resource scarcity. Develop contingency plans to minimize disruptions and protect the company's bottom line. For example, by modeling the impact of wildfires in sourcing regions, Fashion Flair can assess the need for diversifying its supplier base or work in collaboration with its suppliers to take adaptation actions (including vegetation management) that reduce vulnerability to extreme events.

Financial Performance and Risk Management: Quantify the financial impact of sustainability initiatives, including cost savings, revenue growth and risk reduction. Assess the financial implications of climate-related risks and opportunities. For example, by modeling the impact of carbon pricing on product costs, Fashion Flair can develop strategies to mitigate the financial impact and identify potential cost-saving measures.

Materiality Assessment and Disclosure: Utilize UCI to identify and prioritize climate-related financial risks and opportunities for disclosure purposes. Quantify the financial impact of these risks and opportunities to meet regulatory requirements and investor expectations.

Integration with Business Processes: UCI can be integrated into Fashion Flair's product lifecycle management (PLM) system, supply chain management software and financial systems. For example, UCI can provide data on material sustainability and cost to inform product design decisions, or integrate with ERP systems to track the environmental impact of production processes.

By leveraging UCI, Fashion Flair can gain a competitive advantage by making data-driven decisions, mitigating risks and capitalizing on predictive insights within its key markets.

Endnote

1. The Unified Climate Intelligence (UCI) used in this analysis was generated on QuantEarth™, a proprietary analytics engine part of Earthena AI. It used public sources of BHP company data and financial statements to selectively highlight the various aspects of UCI.

CHAPTER 10
UCI FOR TRANSFORMATIVE LEADERSHIP

Summary

In this chapter, we explore the pivotal role of Unified Climate Intelligence (UCI) in redefining corporate leadership. We examine how UCI equips C-suite executives – especially leadership from CEOs and strategic collaboration between Chief Financial Officers (CFOs) and Chief Sustainability Officers (CSOs) – to integrate climate insights deeply into strategy, core operations, financial planning and corporate governance. This accelerating leadership transformation is driven by the need to adapt to the inevitable realities of climate change, which bring both risks and opportunities for value creation.

UCI provides leaders with the necessary tools to make informed decisions that consider both financial metrics and climate action. This strategic integration highlights the evolution of leadership roles within the corporate world, transitioning from current practices to proactive, climate-aware governance that can drive sustainable value creation. For CEOs, this means fostering a climate-literate corporate culture and leading the charge in embedding sustainability into the core business strategy, ensuring that climate considerations are integral to all organizational levels and operations. Climate Gen-AI, powered by UCI, gives everyone in the organization the insights needed to integrate sustainability in their roles and drive more value for the company through climate action.

The collaboration between CFOs and CSOs, facilitated by UCI, exemplifies a shift toward an integrated approach where financial and environmental objectives align. This powerful nexus is crucial in managing the financial implications of climate risks and leveraging opportunities for sustainable growth. UCI is an essential framework for leaders to guide their organizations through the challenges of a climate-impacted future, thereby positioning their companies for competitive advantage in a rapidly evolving market.

The escalating impacts of climate change are not just reshaping the natural environment but are also profoundly transforming the landscape of corporate leadership. As the realities of climate risks and opportunities become increasingly evident, the roles and responsibilities within the organizational C-suite are undergoing a significant evolution. Unified Climate Intelligence (UCI) will play a critical role in redefining how leaders approach their roles and contribute to their organizations as they adapt to change and capture new opportunities.

UCI equips leaders with the necessary tools and insights to make informed decisions that align financial and sustainability objectives. This shift is not merely about adapting to new external pressures but about embracing a leadership style that integrates comprehensive climate awareness into the core DNA of the business. As such, UCI will drive a new era of leadership where executives evolve from current approaches in which climate plays an ancillary role to an approach that integrates climate intelligence into core decision making and views climate as a key driver of future value creation and competitive positioning.

For CEOs in particular, this means championing a climate-literate culture within their organizations. They must ensure that every level of the organization understands the nuances of how climate will impact future business performance and the strategic responses required given its contribution to growth and future financial performance. This involves not only overseeing the integration of climate knowledge across operational processes but also leading by example in prioritizing climate-related objectives. The pivotal role of Chief Financial Officers (CFOs) and Chief Sustainability Officers (CSOs) is also expanding as these officers become crucial in navigating the financial and strategic landscapes shaped by climate considerations. CFOs are increasingly tasked with the financial quantification of climate risks and investments across the enterprise, while CSOs find their roles elevated from managing external reputation to influencing core strategy and embedding environmental objectives deeply within the business.

The shift in the business landscape is driven by the permanent macro trends, requiring a new leadership paradigm that embraces radical transparency, stakeholder management and comprehensive climate disclosures. These trends are making the internal and external views of a company's climate risks and opportunities increasingly visible and financially material, placing the CEO at the centre of orchestrating meaningful change. In this evolving context, the CEO's role extends beyond traditional boundaries to become the linchpin in aligning the organization with a climate-aware strategy that is visible to all stakeholders. This involves not only advocating for and implementing key initiatives but also ensuring that these efforts are transparent and aligned with core strategy.

CEOs play a crucial role in climate-aligning their organizations and leading the charge in closing the gap between perceived and actual climate impacts. As the primary architect of corporate strategy and culture, the CEO's responsibilities include several key activities such as embedding sustainability into governance structures and shaping a vision that helps the organization thrive in a world and economy undergoing climate transformation. There is an ever-increasing expectation for CEOs to be at the forefront of understanding and communicating the company's climate strategy – both internally and externally – and how this contributes to long-term value.

The formation and articulation of a strategic vision that transcends traditional compliance and superficial environmental marketing is a core pillar of the CEO's responsibilities. This means developing and communicating a strategy that focuses on repeatable value creation through climate aligned action. By leveraging UCI, CEOs can integrate real-time data and predictive analytics into their strategy formulation, ensuring that the company's climate actions are not only about meeting regulatory requirements or improving public image but are deeply integrated into the business model and operational processes for true climate-alignment and resilient growth. For instance, a CEO might utilize UCI to identify and develop new market opportunities in emerging green sectors. By analyzing trends and forecasts provided by UCI, a CEO could guide the company to make strategic investments that align the company's growth with global shifts toward sustainability. UCI can also provide a view on the value at risk from future climate impacts if no adaptation measures are taken. This not only positions the company as a leader in the climate economy but also opens up new revenue streams and reduces future losses, enhancing shareholder value over the long term.

Another critical activity in climate leadership is shaping governance structures to support effective climate action as a driver for value creation. This involves embedding climate considerations into the core business strategy and decision making processes. UCI plays a pivotal role here by providing a comprehensive overview of climate risks and opportunities, enabling CEOs to integrate these insights into corporate governance. The CEO can facilitate this integration by establishing clear governance frameworks that assign responsibilities and accountability for climate-related objectives across the organization.

Radical transparency and mandatory climate disclosures have elevated climate considerations to a boardroom imperative. Failure to proactively manage climate risks and opportunities can expose a company to financial and reputational risks, as well as potential litigation. However, UCI provides a powerful capability for the CEO and leadership team to navigate this evolving environment. By offering a unified and auditable data platform, UCI empowers businesses to demonstrate their commitment to climate action with confidence. It fosters trust with investors, regulators and consumers who increasingly value transparency and accountability. Furthermore, UCI systems support C-level executives to see how and where climate action can drive business value. This intelligence can be used to develop key performance indicators (KPIs) or objectives and key results (OKRs) that deliver on organizational climate resilience and focus on value creation, simultaneously providing confidence to external stakeholders such as investors. These KPIs can also be linked to executive compensation, aligning personal incentives with climate goals that deliver financial performance.

The emerging trend of linking executive compensation to climate-related performance metrics signifies an important shift in corporate governance and strategic focus. It introduces a new layer of accountability and is proactive in its pursuit toward climate alignment, incentivizing long-term behavioural change alongside short-term financial performance. However, despite the recognized successes of tying incentives to long-term outcomes, many organizations are still hesitant to integrate such metrics into their executive compensation structures.

Several factors contribute to this reluctance. In the current global economic and geopolitical climate, business leaders are preoccupied with navigating pressing disruptive forces, often prioritizing short-term financial metrics to "keep the ship afloat." Furthermore, ESG regulation remains fragmented, in flux and subject to political polarization – it is a significant challenge for any one individual, department or organization to keep track of. The absence of standardized ESG data collation methods complicates this further, alongside the challenge of establishing and calibrating targets when best practices are scarce and methodologies remain untested. However, by integrating UCI, companies can augment ESG data collation with clear financial metrics, improve target setting accuracy and foster an organizational culture that values sustainable growth alongside financial success. This approach has the potential to not only enhance corporate governance but also positions companies better to meet future regulatory requirements and stakeholder expectations. Importantly, UCI keeps the focus on business value as opposed to box ticking or "feel good" initiatives.

As the primary custodians of corporate vision and strategy, CEOs are uniquely positioned to champion climate literacy within their enterprises.

This role extends beyond mere advocacy; it involves a deep, personal commitment to understanding and integrating climate considerations into the fabric of business decision making. By doing so, CEOs ensure that climate awareness permeates every level of the organization, from the boardroom to the "shop floor." Creating a climate-literate culture is essential for organizations aiming to thrive in the climate economy. This endeavour begins with comprehensive education and training programs that empower employees at all levels with the knowledge and tools to make informed sustainability decisions.

UCI can enhance these initiatives by offering tailored data insights that highlight the specific climate risks and opportunities relevant to various aspects of the business, from supply chain management to product innovation. By integrating UCI into daily decision making processes, CEOs can move beyond high-level models and reports to a more granular and actionable understanding. Applied UCI allows executives to ask and answer critical questions that directly impact their bottom line: What is the climate risk profile of a specific investment opportunity? How can we optimize resource usage to drive both sustainability and cost savings? Which product lines offer the greatest potential for climate value creation? This level of detailed inquiry ensures that climate considerations are not just theoretical but are actively driving the strategic directions and operational decisions of the company.

By embedding climate literacy into the company's core values and operations, CEOs can foster an organizational ethos that prioritizes sustainable practices as a key pillar to business value. UCI facilitates this by enabling seamless integration of climate analytics into daily business processes, decision models, and workflows ensuring that decisions are aligned with the organization's growth and sustainability goals. By effectively communicating the strategic importance of climate initiatives in organizational value creation, the CEO fosters a corporate culture that values lasting sustainability, rather than viewing it as a distraction from core financial objectives. This leadership is crucial in ensuring that the entire organization is aligned and motivated to meet its long-term climate goals, thereby integrating climate considerations into the DNA of the company's strategy and operations for resilience and competitive advantage.

Building on the transformative vision set by the CEO, UCI extends its influence across other pivotal roles within the organization, particularly impacting the potentially powerful collaboration between the CFO and the CSO. By leveraging quantified UCI, the operational synergy between the CFO and the CSO transforms into a powerful nexus, effectively bridging traditional gaps between financial goals and sustainability value drivers, introducing a new layer of intelligence for leaders. This enables a strategic transformation in leadership approaches, enhancing their capability to drive changes, manage risks, and foster competitive advantages under the overarching vision set by the CEO. UCI serves as the digitized central nervous system of this strategy, integrating climate models (physical, transition, nature-based solutions), resource usage data and advanced financial analytics. This comprehensive, 360° view empowers leaders to break down climate risks and opportunities at a granular level – by asset, product, business unit, operating regions or across the entire enterprise – ensuring that climate-related decisions are financially prudent and environmentally sound.

As corporate finance continues to evolve with climate change, the role of the CFO is being reshaped by the increasing pressures on climate related financial impacts. This transformation is driven by several external pressures that necessitate a deep integration of climate considerations into financial strategies:

Regulatory Pressure: With the advent of climate-related financial disclosure regulations, CFOs are required to report on climate risks and opportunities with the same level of rigor and accountability as traditional financial disclosures. This not only involves compliance with emerging standards but also ensuring that such disclosures are comprehensive, reflecting the true financial materiality of climate risks and opportunities in the organization.

Shareholder and Board Pressure: Investors and board members are increasingly demanding that companies align their strategies with climate-conscious objectives. This shift is partly due to a growing recognition of the financial risks and opportunities that climate change presents. Additionally, legal liabilities are now a concern for board members and executives who must understand and act upon future climate-related impacts to safeguard shareholder value and meet fiduciary duties.

These pressures underscore the necessity for CFOs to enhance their understanding of how climate factors affect the business, both currently and in the future, and to articulate this impact through clear financial projections and strategic initiatives. As part of their core competencies, CFOs are finding that integrating UCI is not just an option but an inevitable shift that enhances their ability to manage financial risks and identify new growth opportunities in a climate-conscious market.

For pioneering CFOs, UCI is a powerful strategic lever to drive substantial future value and can elevate several aspects of their role. For example, UCI allows CFOs to integrate climate risks and opportunities directly into the financial architecture of their organizations. This integration goes beyond mere compliance, enabling CFOs to apply best-in-class scientific data and analytics to refine core financial statements – such as profit and loss statements, balance sheets, and cash flow projections. By making climate considerations a routine part of financial reporting, CFOs can provide a more accurate picture of future financial scenarios impacted by climate variables. This rigorous approach helps in articulating the financial materiality of climate impacts, facilitating strategic planning and communication with investors and stakeholders.

With UCI, CFOs are equipped to embed climate-related insights into critical financial decisions including capital allocation, investment strategies, research and development budgeting and obtaining favourable financing terms. This strategic integration ensures that financial decisions are informed by comprehensive climate risk assessments and potential opportunities, thus aligning the company's financial planning with long-term sustainability goals. For example, UCI can identify how shifts in climate policy might impact energy costs or supply chain logistics, allowing CFOs to make proactive adjustments to the company's investment strategies or operational plans.

Finally, beyond compliance and strategic alignment, UCI enables CFOs to identify efficiencies and cost-saving opportunities that align sustainability with financial performance. By analyzing data on resource use and emissions, CFOs can pinpoint areas where improvements can lead to significant cost reductions while also advancing the company's sustainability objectives. These might include enhancing energy efficiency, reducing waste or optimizing logistics to lower carbon footprints and future carbon costs.

The proactive use of UCI transforms the CFO's role into a dynamic enabler of climate-resilient growth. It allows the finance function to anticipate and mitigate financial risks from climate change proactively, while also capitalizing on new market opportunities created by the evolving regulatory and economic landscape. This dual focus not only secures the long-term financial health of the company but also positions it as a leader in the emerging climate economy.

Historically, sustainability was perceived as a marketing function, or perhaps a standalone facet focused on discrete initiatives that could be tracked and reported as part of a larger initiative, such as reputation management and external stakeholder engagement. Today however, the CSO is emerging as a strategic partner within the C-suite and to the board, crucially involved in core business operations. This shift demands that CSOs and their teams effectively translate climate data and sustainability initiatives into core business strategies that are financially quantified. With UCI, sustainability is integrated into the decision making fabric of the organization, enabling departments across the enterprise to utilize this intelligence to meet their specific objectives and contribute meaningfully to overall business goals. Thus, UCI not only enhances the strategic function of the CSO but also aligns sustainability with the financial and operational priorities of the company.

UCI transforms the role of the CSO, arming them with the analytical capabilities needed to rigorously identify and prioritize actions that drive business value and mitigate risks. It enables CSOs to articulate the financial implications of sustainability initiatives clearly and compellingly, crucially highlighting the significant costs of inaction alongside the benefits of proactive measures. As previously discussed, this increasingly essential insight allows the C-suite and board to make well-informed, data-driven decisions that align environmental responsibility with business objectives. By presenting a clear financial narrative of action versus inaction, UCI strengthens the CSO's role as a strategic partner, ensuring that sustainability is woven into the fabric of corporate strategy and decision making.

Below, we explore how UCI can be seamlessly integrated into the financial and sustainability frameworks of an organization. Using a practical example this inset turns climate data into a common currency that drives value creation and strategic decision making across financial and sustainability teams.

Example 10.1 Strategic Deployment of Unified Climate Intelligence (UCI) into Financial and Sustainability Architecture

Integrating UCI into the financial and sustainability architecture of an enterprise is not just about compliance or risk management, but about embedding a new dimension of intelligence into the core of business decision making. This integration allows companies to navigate the climate economy with a strategic edge, making informed decisions that align with climate resilience, financial performance goals and broader sustainability objectives.

Linking UCI to Financial and Sustainability Goals:

- Use UCI to enhance financial reporting by linking climate risks and opportunities directly to financial statements, such as P&L reporting. This approach highlights the financial benefits and costs of sustainability initiatives, such as resource efficiency and waste reduction, within operational budgets.

- Apply UCI for asset valuation adjustments on the balance sheet, assessing the full spectrum of climate risks. This includes modifying property values based on climate vulnerability and adjusting investment valuations to reflect resource scarcity and circular economy opportunities.

Incorporation into Financial and Sustainability KPIs:

- Develop combined KPIs that blend traditional financial metrics with climate intelligence and sustainability factors, such as "Climate-adjusted Return on Investment" or "Climate Risk-adjusted Profit Margins."
- Use UCI to refine metrics like Earnings before Income Tax (EBIT), factoring in potential costs or savings from climate strategies and resource efficiency initiatives as well as revenue enhancements from strategic opportunities.

Climate Disclosures and Materiality:

- Leverage UCI to ensure comprehensiveness and financial materiality in climate disclosures. Identify and quantify climate-related risks and opportunities that have a significant financial impact on the business, aligning with evolving regulatory requirements and stakeholder demands for transparency.

Margin Analysis and IRR with Sustainability Considerations:

- Apply UCI data to conduct a nuanced margin analysis, considering how climate risks and opportunities, alongside resource efficiency and potential circular economy opportunities, could affect different product lines or business units. This can be done for a range of potential future climate scenarios to understand sensitivity and exposure to a range of future policy (e.g., carbon price effects) and physical risk (e.g., increasing intensity and frequency of extreme events) scenarios.
- Calculate internal rate of return (IRR) for climate-related investments and resource efficiency initiatives, demonstrating

(*continued*)

how proactive climate and sustainability strategies can yield financial returns.

Hurdle Rate Adjustments with Sustainability Integration:

- Adjust hurdle rates for new investments by incorporating climate risk assessments and sustainability benefits, ensuring that new projects are resilient, financially viable and contribute to the company's sustainability goals.

Linking UCI to Financial Planning and Budgeting with Sustainability:

- Embed climate risk, opportunity costs and sustainability potential into annual financial planning and budgeting cycles.
- Allocate budgets based on UCI insights, prioritizing areas with the highest combined climate-related financial impact and sustainability benefits.

UCI in Capital Allocation Decisions with Sustainability:

- Inform capital allocation decisions with UCI insights, directing funds toward projects and initiatives that offer the best climate-aligned value creation opportunities while considering resource efficiency and circular economy potential.

Creating Financial and Sustainability Artifacts for Decision Making:

- Develop financial models and artifacts that use UCI data to project long-term financial impacts of climate and sustainability strategies.
- Enable scenario analysis to visualize financial and sustainability outcomes under different climate futures.

Enhancing Investor Relations and Reporting:

- Utilize UCI to provide detailed, quantified climate-related disclosures to investors, enhancing transparency and trust.
- Report on climate-related financial performance and sustainability achievements in shareholder communications, annual reports and investor briefings.

A further area to consider of UCI's role in transforming leadership for climate value creation is its horizontal capability. This aspect is crucial in achieving true organizational climate alignment, which requires embedding UCI across all functions – from operations and product development to finance and investment planning. This ensures that climate considerations are not an afterthought, but a core competency woven into the fabric of every strategic and operational decision. By integrating UCI and powered by Climate Gen-AI, C-suite leaders can leverage actionable intelligence to address critical climate challenges within their specific areas of responsibility. Here are some examples:

The Chief Marketing Officer (CMO):

> "How can we leverage UCI to identify emerging sustainability trends and translate them into high-growth product or service opportunities?"
>
> "What is the potential financial impact of shifting consumer preferences toward sustainable alternatives within our product categories, and how can we adjust our marketing strategies to capitalize on this trend?"
>
> "Where can we optimize our marketing spend to reach climate-conscious consumers most effectively across different geographic regions?"

The Chief Operating Officer (COO):

> "What is the potential financial impact of climate-induced disruptions on our supply chain over the next decade, and how can we use UCI to identify and prioritize the most cost-effective mitigation strategies?"
>
> "By leveraging UCI, can we identify opportunities to improve resource efficiency and waste reduction across our operations, leading to both environmental and cost-saving benefits?"

"Where can we strategically invest in green infrastructure upgrades to enhance operational resilience against climate risks, such as extreme weather events, and what is the projected return on investment (ROI)?"

Finally, in our second example, we demonstrate how to integrate UCI into your organization, fostering a culture of climate accountability and enterprise-wide alignment using an example road map.

Example 10.2 Unleashing the Power of UCI: A Road Map for Enterprise-Wide Climate Action

A Structured Approach to UCI Integration:

1. *Building a Shared Understanding: Educate all departments on UCI's capabilities, emphasizing its ability to:*
    - Quantify climate risks and opportunities across physical, transition and nature-based factors.
    - Model different climate scenarios to assess potential impacts and inform strategic decision making.
    - Analyze climate data alongside net zero goals, circularity potential, decarbonization pathways and resource availability (e.g., renewable energy supply).
2. *Identifying Departmental Climate Touchpoints:*
    - Supply Chain Management: Integrate UCI to assess supplier vulnerabilities and develop alternative sourcing strategies that enhance resilience.
    - Operations: Use UCI to evaluate climate risk impacts on operations and strategize adaptive measures like infrastructure reinforcements or process optimizations.

- Finance: Implement UCI in financial planning and risk assessment, factoring climate risks and opportunities into investment decisions and capital allocation.
- Marketing and Customer Relations: Leverage UCI data to understand customer sentiment on climate issues and develop marketing strategies that highlight the company's commitment to sustainability.

3. *Seamless Integration:*
   - Develop user-friendly interfaces or dashboards where UCI insights are easily accessible to relevant departments.
   - Ensure smooth data integration with existing decision making tools and platforms, including relevant workflows and business processes.
4. *Building Expertise:*
   - Conduct training sessions for each department, focusing on interpreting and using UCI data effectively, with an emphasis on AI-assisted expertise and advanced analytics techniques, particularly Gen-AI.
   - Develop a resource hub with guides, case studies and best practices on integrating UCI into departmental workflows.
5. *Goal Setting and Measurement:*
   - Establish clear, measurable climate-related goals for each department based on UCI insights, with a clear link to departmental KPIs. Examples include reducing supply chain carbon footprint or improving operational resilience.
   - Regularly review these goals and the effectiveness of implemented strategies.

(*continued*)

6. *Cross-Departmental Collaboration:*
   - Foster collaboration between departments to tackle complex climate challenges.
   - Leverage UCI powered Gen-AI to enhance collaboration and knowledge sharing across the organization.
7. *Continuous Learning and Improvement:*
   - Continuously monitor the effectiveness of UCI integration across departments.
   - Be adaptable and ready to tweak strategies based on new insights or evolving climate conditions.
   - Establish feedback mechanisms where departments can report on the utility of UCI and suggest improvements.
   - Use this feedback to refine UCI integration and ensure it remains a dynamic and valuable asset to the organization. Climate literacy fostered through this process empowers more informed decision making and long-term value creation and increasingly we see Gen-AI playing a significant role in provisioning role-based climate intelligence and training.

Consider this road map as a living document that evolves alongside the organization's growing understanding of climate intelligence and its applications. Successful UCI integration is a journey of continuous learning, adaptation and innovation. By empowering leaders across the organization and fostering a culture of collaboration, UCI becomes a catalyst for transformative climate action, driving long-term business success and a more sustainable future.

The role of UCI in transforming corporate leadership and fostering a climate-aware culture within organizations cannot be overstated. By embedding UCI into the strategic core of business operations, leaders are empowered to not only manage and mitigate risks but also to seize the opportunities that a transition to a sustainable economy presents. This chapter underscores how UCI builds competitive advantage, advocating for its embrace as a path to transformative leadership and sustainable value creation. Through UCI, augmented with Gen-AI, companies are equipped to lead in a rapidly evolving, climate-conscious market, ensuring that their strategic initiatives are aligned with both current needs and future sustainability goals.

CHAPTER 11
CATALYZING UCI THROUGH FINANCIAL SERVICES

Summary

This chapter underscores the pivotal role of financial services in harnessing Unified Climate Intelligence (UCI) to steer global financial systems toward robust climate alignment. It outlines the need for climate intelligence to be integrated more systematically in our financial system and showcases how UCI can reshape financial practices across various sectors including insurance, credit, private equity, asset management and central bank stress testing. By integrating comprehensive climate data, these financial sectors can not only understand financial materiality, but also adapt and capitalize on the opportunities presented by the emerging climate economy, stimulating economic growth and job creation.

The transformative potential of UCI in refining risk assessments, enhancing investment strategies and influencing policy decisions is clear. As financial services embrace and integrate climate intelligence, they become crucial catalysts in the global effort against climate change. We demonstrate that with the right tools and strategies, financial markets can lead the charge toward sustainable economic development, aligning financial flows with the urgent needs of our planet for resilience and sustainability.

Climate change presents not only environmental opportunities and threats, but also new and complex financial challenges. The role of financial services in addressing these issues is becoming increasingly pivotal. As central pillars of the global financial system, capital markets influence economic growth and technological innovation by mobilizing and allocating vast financial resources. They are instrumental in financing large-scale transitions to low-carbon, sustainable practices across industries globally. However, despite their significant influence, capital markets have often lagged in integrating climate considerations into their financial models, transactions, risk and portfolio management practices and decision making processes comprehensively. Traditional financial models typically overlook the long-term impacts and risks associated with climate change. Historical climate stability has meant that capital markets and associated sectors have not had to model future climate volatility. This leads to a significant underestimation of potential liabilities and mispriced assets, undermining the ability of financial markets to protect against climate-induced economic shocks and to capitalize on opportunities arising from the transition to sustainable practices. In a world of increasing climate volatility, changing regulations and dramatic technological shifts, the tools of the past are no longer fit for purpose.

The introduction of Unified Climate Intelligence (UCI) marks a transformative shift in how financial markets can comprehensively address climate change. UCI offers a holistic approach to understanding and acting on climate data, enhancing the financial sector's capacity to assess and manage climate-related risks and to drive investments toward more sustainable and resilient economic activities. Fundamentally, UCI enables financial services to analyze future looking climate drivers and incorporate multiple layers of financial modeling. This includes the impacts of climate at the macro and micro levels. At the macro level, integrating UCI provides the ability to inform the aggregate impacts of multiple climate drivers on the economy at large. This includes how physical and transition risks will affect drivers such as GDP growth, inflation, productivity effects, resource pricing and more. These are important inputs and assumptions that drive macroeconomic models and ultimately relate to factors such as equity prices, investment strategies and asset allocations.

Multiple scenarios need to and should be modelled to understand the possible futures and the impacts they will have on the macro environment. UCI also provides insight at the micro level. This includes the effect of climate on individual firms or households. Using UCI to understand future impacts on the future cash flows of a specific company or on the property value of a home, enables financial services to more adequately assess the risks and opportunities at the micro level, informing decisions in areas such as corporate lending, private equity, and mortgage portfolios as well as insurance and reinsurance pricing.

Capital markets have the power to drive the mobilization and allocation of vast financial resources. By leveraging UCI, investors, financial institutions and policymakers can accelerate climate action by enhancing the accuracy of risk assessments, refining investment portfolios to favor low-carbon solutions, and developing financial products that directly support climate resilience and adaptation efforts. There are already early signs that integrating climate into the financial system is not just beneficial but necessary. Green bonds, where proceeds are exclusively applied to finance or refinance new or existing climate-related or environmental projects, are becoming increasingly popular. Investment firms are using climate intelligence to re-evaluate and adjust their asset allocations, moving away from sectors heavily reliant on fossil fuels toward industries that are actively reducing carbon footprints. The influence of integrated UCI extends deeply into specific sectors such as insurance, credit and private equity, where the precise management of climate risks is becoming increasingly critical for stability and growth. And, as we discussed in Chapter 5, governments and regulatory bodies are shaping policies that encourage financial markets to incorporate climate considerations into their decision making frameworks.

The insurance and reinsurance industries have and will be deeply impacted by climate change, necessitating a fundamental transformation in how risks are assessed, managed and priced. Traditional models, such as property insurance and catastrophe modeling, are proving inadequate in the face of the escalating frequency and severity of climate-related events. These models often fail to capture the full spectrum of impacts from climate change and policies. For example, the property insurance market is reliant on being able to model – with a high degree of confidence – the likelihood and impacts of extreme events based on patterns of the past. Typical catastrophe models are based on historical, backward- looking data and do not effectively account for forward-looking climate projections. This leads to significant gaps in risk assessment, financial planning and product offering; e.g., the escalating frequency and severity of wildfires in California have led insurers to limit coverage or significantly increase premiums. Similarly, coastal properties face rising insurance costs due to increased hurricane and flood risks.

The challenges in the insurance sector also have knock on effects in other areas of the financial system. For example, most mortgages – typically with a 15–30 year term – require insurance. However, property insurance is priced and offered annually with no guarantee that insurance will be available for the lifetime of the mortgage (as required by the banks) or for the lifetime of the property (as would be ideal for the property owner). When reinsurers and insurers decide that they are no longer comfortable taking the risks, it leaves the rest of the financial system built on that risk insurance and the property owners exposed. In many places, the government is stepping in to fill this gap, but how long and to what extent can that continue? In other cases, there is a significant increase in premiums affecting affordability for homeowners and increasing probability of defaults. Ultimately the underlying asset value will have to be repriced and current mispricing of the value can have a cascading effect in the future as we have seen with other housing related financial crises.

Given these new realities, insurers and reinsurers are under increasing pressure to innovate and adapt their practices to better reflect the evolving nature of climate threats. It is here that UCI becomes crucial, offering new ways to integrate comprehensive climate data and predictive analytics into the heart of insurance operations. By doing so, UCI enables — insurers and reinsurers to develop more dynamic models and to innovate their product offerings in order to keep pace with the rapid changes in climate risk landscapes.

To stay competitive and effective amidst these challenges, the insurance and reinsurance industries are turning to innovative solutions like parametric insurance. This recent approach differs from traditional models by utilizing predefined climate events to trigger immediate financial payouts. For example, payouts are automatically disbursed when a hurricane reaches a specified wind speed or a river exceeds a certain water level. This model drastically reduces the administrative burden of claims processing and provides quick financial relief, making it an invaluable tool in the immediate aftermath of a disaster. Valued at $11.7 billion in 2021 and expected to reach $29.3 billion by 2031, parametric insurance exemplifies the shift toward more efficient risk management tools that cater to the urgent needs of climate risk management. The rapid growth of this market underscores the increasing demand for financial products that can efficiently address the challenges posed by climate change. The straightforward criteria and rapid payout mechanisms of parametric insurance expand its accessibility, making it especially beneficial in regions frequently hit by climate-induced disasters. The integration of UCI with parametric insurance enhances the precision of the parameters used in these policies, ensuring that they are closely aligned with the most current scientific understanding of climate variability and change. This alignment not only improves the responsiveness of insurance products but also broadens their impact, helping to buffer financial impacts in the face of climate uncertainty and assist with managing and mitigating climate risk.

The integration of real-time data from satellite imagery, weather stations and advanced climate models by UCI can also enable insurers to develop dynamic pricing models that more accurately reflect the increased risk and potential damage from extreme weather events such as hurricanes, floods and wildfires. This enables them to offer innovative and customized insurance products with premiums that adjust based on updated risk assessments factoring in forward looking climate aware models.

There is a significant opportunity to promote climate resilience through strategic use of insurance premiums. This would require insurers and reinsurers to not only look at the exposure a property or business operation has to volatility, but also its adaptive capacity. For example, if a property owner, business or community takes measures to increase the threshold at which they are exposed to an extreme event or increase the coping and recovery capacity to an event, they have reduced their vulnerability, thereby reducing their risk profile on damage and disruption. Just as health insurance companies update their models based on ongoing patient and health data, insurers have an opportunity to monitor and incorporate data on adaptation actions and adjust premiums accordingly. Using UCI to model changes to insurance premiums by integrating adaptation and mitigation actions has the potential to create financial incentives to drive greater climate resilience at the individual and community level.

UCI also assesses the financial implications of ecosystem damage, such as biodiversity loss and disruptions to ecosystem services, which can affect everything from natural disaster protection to water quality and crop yields. Insurers can use this data to price policies that cover ecological restoration or compensate for the loss of ecosystem services. As such, UCI supports the creation of new insurance products aimed at funding ecosystem restoration projects, promoting investments in nature-based solutions that bolster community and business resilience.

The role of climate intelligence in insurance will continue to grow. Emerging trends include the development of more sophisticated models that incorporate a wider range of data inputs, including economic and social parameters alongside environmental factors. Artificial intelligence and machine learning are enhancing the predictive accuracy of climate models, leading to more effective risk mitigation possibilities. As climate science advances and provides deeper insights into the long-term risks and opportunities of climate change, insurance products should evolve to manage these more effectively.

Climate risk and opportunity in credit analysis and decision making can significantly affect the creditworthiness of borrowers and the performance of loans. The integration of climate-related factors into credit assessment is becoming increasingly important as these risks – from physical impacts like storms, wildfires and floods to transitional risks associated with shifts toward a low-carbon economy – introduce layers of financial uncertainty. For example, properties or businesses located in areas prone to extreme weather events may face higher costs for recovery and insurance, impacting their ability to fulfil credit obligations. Similarly, companies in industries reliant on fossil fuels may encounter depreciating asset values and profitability challenges due to policy changes, such as carbon pricing and emissions regulations. By assessing how these climate risks and opportunities could affect a borrower's cash flow, collateral value and overall financial health, lenders can make more informed decisions that enhance the resilience of their portfolios and encourage borrowers to adopt sustainable practices to maintain or improve their creditworthiness.

UCI can enhance the process of credit analysis by integrating climate risk assessments directly into the financial decision making frameworks. In the case of physical risk, this integration involves utilizing advanced climate forecasting and risk modeling tools that incorporate a wide range of data points, from historical weather patterns to future climate change projections. Credit analysts can use this comprehensive data to create more robust risk assessment models that factor in the probability and potential impact of physical climate events on borrowers' financial stability. Specifically, UCI can be integrated into core credit processes such as probability of default models (analyzing likelihood that a borrower will default on their loan obligations within a specific time period) and loss given default models (analyzing the amount of money the bank will lose when a default occurs). Integrating UCI gives credit analysts the tools to evaluate the potential impacts of physical damages and disruptions that might affect the collateral or business operations underpinning loans. For instance, when assessing mortgages with long durations, such as 15 to 30-year loans, UCI can provide lenders with predictive insights on the evolving climate risks associated with a property's location. By mapping out areas likely to be impacted by sea-level rise or increased frequency of extreme weather, UCI enables banks to adjust their loan terms, set appropriate interest rates, or require additional safeguards like insurance or adaptation actions to manage these risks. This proactive approach helps banks safeguard their loan book against future climate-induced devaluations and ensures borrowers are better prepared for potential challenges.

For transition risk, UCI offers detailed scenario analysis tools that predict how different regulatory and market developments could impact specific industries. Analysts can use these tools to evaluate the long-term sustainability of business models and adjust their credit strategies accordingly. This may include reducing exposure to high-carbon emitting sectors, diversifying portfolios or increasing investments in companies leading the transition to sustainability. For example, investments in coal-fired power plants are at high risk of becoming stranded assets due to stringent emissions regulations and shifting energy markets favouring renewables. Here, UCI aids credit analysts in identifying such risks by providing insights into regulatory changes, technological advancements and market trends that may affect the viability of certain sectors or assets.

To drive climate aligned decision making, UCI should be integrated into multiple parts of the credit process. This includes: (i) prospecting and origination to target appropriate sectors, regions, companies and/or assets; (ii) underwriting and risk pricing to assess the risk and price it appropriately; (iii) portfolio management and risk management to hedge the risk and assess the risk that bank is carrying on its books, including stress testing its credit portfolios; and (iv) opportunities for new products and innovation.

New products and innovation will be critical in accelerating climate action. We have already seen increased credit products (e.g., green bonds) to finance areas such as the energy transition. There is room for far more innovation, and substantially greater amounts of capital will be required to climate align and future proof the economy. For example, financial institutions have the opportunity to explore adaptation and resilience financing, especially if private finance can be blended with public capital to support the large scale infrastructure investments. UCI that informs credit modeling can also be used to support informed infrastructure planning, building codes, critical failure thresholds and more. It is in everyone's interests to ensure that the financing being provided is going toward creating more resilient assets and operations.

Climate change has also been moving to the front of the agenda at central banks and other financial supervisory bodies. In particular, they have started assessing the resilience of the financial system to increasing climate volatility and have been mandating banks to stress test their exposure to climate related risks. As the regulatory heartbeat of the financial sector, central banks utilize stress testing to set standards and expectations for how financial institutions should manage their exposure to climate risks. This systemic approach not only assesses the resilience of individual entities but also examines the interconnected vulnerabilities that might cascade through the financial system, and is designed to ensure that banks maintain sufficient capital to withstand significant economic disruptions. Traditionally, stress tests have focused on economic downturns and financial market shocks, but the escalating impacts of climate change necessitate a broader scope that includes environmental risks. Scenario analysis is an important component of stress testing, whereby banks assess their exposure and vulnerabilities under different future climate scenarios. Central banks have been acutely aware of climate change as a systemic financial risk for a long while, and have been instrumental in driving the global push for climate-related financial disclosures.

UCI will profoundly transform this process by providing more precise and comprehensive data on climate-related risks and opportunities. By incorporating scenarios that account for physical, nature and transition risks associated with climate change, UCI equips banks with the tools to conduct more targeted and effective stress tests. This integration helps identify potential vulnerabilities in the financial system that could be triggered by climate-related events, thus guiding banks in developing robust mitigation and adaptation strategies. These might include developing plans that focus on the geographical diversification of assets and implementing infrastructural resilience measures against physical risk. In the case of transition risk, central banks can use UCI to evaluate how asset valuations, credit risks and the overall financial health of institutions are affected. The integration of these insights into stress testing helps banks adjust their portfolios to mitigate risks associated with policy changes, such as carbon pricing or subsidies for renewable energy.

Due to the interconnectedness of global markets and value chains, climate change has the potential to exacerbate systemic risks and the stability of the entire financial system. UCI provides a macroscopic view of potential systemic disruptions, including the cascading effects of climate risks across sectors. For example, a major flood could not only impact insurers and reinsurers but also the banks that finance homeowners and businesses in the affected area. UCI tools help in mapping these interdependencies and developing comprehensive stress testing scenarios that include cross-sectoral and networked impacts, allowing banks to evaluate strategies for systemic and compound risk management, such as enforcing higher capital buffers and promoting liquidity support measures, safeguarding against climate-induced disruption.

To mitigate systemic risks effectively, central banks around the world are implementing various strategies, primarily through enhanced stress testing scenarios and macroprudential policies. Macroprudential policies focus on preventing financial crises by managing the entire financial system's resilience rather than just individual institutions. Successful integration of these strategies into central bank operations often involves collaboration with international financial organizations to establish best practices and standardized approaches, e.g., the Network for Greening the Financial System (NGFS), which provides guidelines and recommendations for incorporating climate risk considerations into central banking practice.

The integration of UCI into macroprudential policies allows for more precise identification and quantification of risks, helping to tailor regulatory buffers and capital requirements to risk levels. It also supports the standardization of risk assessment methods across different jurisdictions within the NGFS, fostering a more cohesive global approach to financial stability. Enhanced scenario analysis, facilitated by UCI, offers richer, more detailed environmental conditions for stress testing, enabling regulators and financial institutions to better understand potential climate impacts. By improving data sharing among NGFS members through advanced data capabilities, UCI promotes more efficient knowledge exchange, enhancing global efforts to green the financial system.

Climate stress testing is in its early stages and still primarily being used as an exercise to start integrating climate risk into frameworks, models and practices with no direct capital requirements. There will inevitably be more stringent requirements in how climate related systemic risks are managed including capital requirement mandates and integration into risk management frameworks of banks and other financial institutions. UCI will play a critical role in how these risks are managed in the future and enables banks to model top-down (e.g., climate impacts on economy or sector-wide drivers) as well as bottom-up (e.g., asset-level insights on specific properties and businesses) impacts more effectively.

Private equity and asset management play crucial roles in providing capital to industries and organizations, from start-ups needing growth capital to established companies looking for restructuring or expansion opportunities. Strategies employed in private equity investments include leveraged buyouts, venture capital, growth capital, distressed investments, and mezzanine capital. Asset management, on the other hand, involves managing investments on behalf of others. This includes a wide range of assets such as stocks, bonds, real estate and more, held through mutual funds, hedge funds and investment trusts. Asset managers make investment decisions to meet specified investment goals for the benefit of their clients.

In private equity, the investment strategy is typically characterized by active ownership. This means investors not only provide capital but also bring in management expertise, industry connections and operational improvements to drive value creation within the portfolio companies they acquire. The ultimate goal is to enhance the companies' value over a period of a few years, before exiting the investment at a significant profit. Asset management strategies vary widely depending on the fund's investment mandate and the risk tolerance of its clients. Strategies can range from conservative – focusing on income and preservation of capital through investments in government bonds or blue-chip stocks – to aggressive strategies that aim for high returns by investing in more volatile assets such as emerging market equities or high-yield bonds. Sustainability and integration of ESG factors are increasingly becoming part of asset managers' strategies as investors become more conscious of the impact of their investments; however, they do not provide the strategic insights that UCI can deliver.

The integration of UCI into the investment processes, such as due diligence and valuations allows private equity firms and asset managers to make more informed decisions, thereby enhancing the resilience of their portfolios to climate-related financial impacts and positioning them for better long-term performance in an increasingly uncertain global market. For private equity, where the investment horizon often spans several years, the use of UCI is invaluable in forecasting long-term climate impacts and aligning investment strategies with anticipated market conditions. It aids in selecting companies that are not only resilient to climate risks but are also well-positioned to capitalize on opportunities arising from the transition to a sustainable economy. In asset management, UCI can enhance portfolio diversification strategies by identifying sectors and geographies with lower climate risk profiles or higher potential for growth in the emerging climate economy.

Mitigating climate risk and identifying opportunity is becoming an indispensable strategy for sustaining and enhancing investment returns, not least by having the ability to engage with portfolio companies on climate issues. By leveraging UCI, asset managers can guide companies in implementing more robust climate governance frameworks, improving their carbon footprint, and developing sustainable business models. This not only helps reduce climate-related risks but also enhances the company's market competitiveness and compliance with emerging regulations.

As global implications of climate change manifest, the demand for investors to understand financial materiality will inevitably rise given these new risks. In tandem, investors are increasingly looking for opportunities that not only provide financial returns but also contribute positively to environmental and social outcomes. The integration of UCI into investment strategies will become a standard practice, driving innovation in financial products and investment models that directly address the challenges and opportunities presented by climate change. Mitigating climate risk will become synonymous with securing future asset value and capitalizing on new growth opportunities.

As banks, insurers, asset managers, equity analysts and others start incorporating UCI in their models and processes, there will inevitably be pressure on CFOs, CEOs and board directors to proactively manage their climate related financial performance. The availability and granularity of climate intelligence will only increase and with this information in the hands of capital markets and financial institutions, there will be no place for companies to hide. Leaders should proactively leverage UCI to make decisions that optimize their climate related financial performance rather than be blindsided by their lenders, investors, insurers or supply chain partners.

By facilitating a deeper understanding and better pricing of climate risks, UCI also paves the way for significant economic opportunities within the public realm. For example, the current financial requirements for adaptation are significantly underserved by international public finance flows, with estimated costs for adaptation being nearly 20 times higher than the funds currently available. UCI can play a transformative role in addressing this gap by enhancing financial confidence and streamlining approaches, thus unlocking the flow of funds necessary for climate action. It can enable the development of financial instruments that more accurately assess and protect against climate risks, promoting not only sustainable investment but also supporting the development and deployment of innovative financing solutions like the Loss and Damage Fund, which aims to assist those most affected by climate change impacts.

The application of UCI in the public sector might also guide the creation of financial instruments and products (e.g., first-loss guarantees) that ensure essential safety nets are in place, especially for vulnerable communities at high risk from climate impacts. The necessity for public and private partnerships in financing climate adaptation efforts is increasingly evident, as market mechanisms alone will not sufficiently cover or insure against all climate risks. This integrated approach encourages a more resilient and proactive stance in financial planning and risk management, and would aim to bridge the adaptation finance gap effectively and sustainably.

The heightened focus on climate opportunity and risk is driving a fundamental transformation in capital markets. Advanced analytics, including climate-focused data analytics and machine learning models, are becoming essential tools for assessing and pricing risk. As a result, financial decisions are increasingly aligned with climate considerations. Financial institutions can now ask questions that they weren't even a few years ago, and UCI gives them the ability to begin answering these questions. Asset managers can ask what the sources of climate risk are in their mortgage portfolio or how they can segment a diverse portfolio by risk exposure. Investment bankers and private equity analysts can ask what the financial impacts on their clients and companies will be from taking specific mitigation and adaptation actions or how to incorporate climate into forward-looking financial statements and valuations. Banks and credit analysts can ask about the climate related correlations between their credit and mortgage portfolios or how to stress test their credit portfolios against multiple future scenarios and tail risk events. Insurers and reinsurers can ask how forward-looking climate models can be used to price more effectively while also looking to maintain affordability or how actions taken to reduce vulnerabilities can be priced to incentivize actions that build resilience.

Moreover, the era of radical transparency demands greater scrutiny of financing activities, pushing financial institutions toward climate alignment. Ultimately, climate risk is being recognized as financial risk, and funding decisions will increasingly favor climate-aligned assets and businesses.

Climate-aligning finance hinges on our ability to integrate comprehensive climate intelligence into every facet of economic planning and decision making. As climate change continues to pose significant global challenges, the financial sector's response will be crucial in shaping a sustainable future. By harnessing the power of UCI, financial markets can not only protect against climate-induced disruptions but also drive the innovation and transformation needed for a resilient global economy. The ongoing developments in UCI and its application across finance underscore the dynamic evolution needed in response to the urgent demands of climate change, setting a new standard for how the financial world operates within the limits of our planet's natural systems.

CHAPTER 12
UCI AS A CATALYST FOR BROADER SOCIETAL CHANGE

Summary

This chapter explores the transformative potential of Unified Climate Intelligence (UCI) in addressing the multifaceted challenges of climate change. The application of UCI extends far beyond organizational benefits to significantly influence societal progress and inform global climate policies. By integrating a broad spectrum of climate data – from physical, nature and transition risks to socio-economic impacts – UCI provides a holistic view that enhances the formulation and execution of climate strategies, ensuring they are comprehensive, sustainable and aligned with both local needs and global objectives.

UCI plays a crucial role in facilitating regional adaptation and progressive climate policies. It empowers communities, cities and countries by providing the essential insights needed to make informed decisions. UCI can be a powerful tool in the implementation of effective and equitable climate action across stakeholders.

As climate challenges intensify, the chapter underscores the urgent need for the widespread adoption of UCI to navigate these complexities effectively. It calls for a unified effort among governments, civil society, communities and enterprises to leverage UCI, including Gen-AI, in creating resilient value, driving a collective movement toward comprehensive climate responsibility and sustainable development.

As outlined throughout this book, Unified Climate Intelligence (UCI) provides numerous organizational benefits; however, its influence extends far beyond the confines of individual companies. Climate change has transformed from an unpredictable force to a more knowable and quantifiable risk that affects us all globally, from nations to individual citizens. Today, the stakes are higher than ever; our climate systems have already locked in significant risks for decades to come, while other challenges – such as nature degradation, transition pressures and escalating carbon emissions – continue to accelerate.

With our advancements in scientific understanding, the proliferation of digital infrastructure and an increasing collective expertise, we already possess the tools necessary to deploy UCI on a global scale. The time to harness these capabilities to adapt to and mitigate climate impacts simultaneously is now. UCI can empower not only countries but also cities, communities and individuals with the insights and tools needed to make informed decisions, driving societal progress and shaping effective global climate policies. Climate scientists are sounding the alarm on the negative impacts of climate change, but we are not powerless. We have deployable technologies with falling costs, creating a critical window of opportunity to act. So, the universal deployment of UCI represents a sensible strategy to navigate the complexities of climate change, ensuring a sustainable future for the next generations.

UCI transcends traditional data analysis by incorporating a wide range of climate factors – from physical risks to socio-economic impacts – offering a nuanced perspective that enriches the formulation and execution of policies. Current initiatives such as Nationally Determined Contributions (NDCs), while critical, often narrowly focus on CO_2 emissions. UCI's broader lens not only facilitates a more interconnected approach by considering a range of environmental factors, but it also integrates insights across adaptation, mitigation and nature-based solutions (NbS).

This holistic view enables a more effective response to the complexities of global warming by ensuring that policies and actions are not blinkered or siloed. Instead, UCI supports the development of policies that are both environmentally sound and economically viable, making climate action comprehensive and sustainable. Importantly, this unified perspective enables decision makers to craft more effective strategies that encompass the interconnectedness of climate challenges, ensuring that actions in one area reinforce, rather than undermine, efforts in another.

Furthermore, UCI enhances public-private partnerships by providing a transparent framework for the targeted allocation of funds dedicated to climate interventions. UCI provides deep insights into the financial, social and environmental cost of inaction and the quantified benefits of action. This capability is crucial for ensuring that investments are impactful and address the most critical needs without unintended negative consequences. Leveraging UCI insights for climate resilience, adaptation and mitigation projects can boost stakeholder confidence, streamline processes and unlock vital funds for climate action – potentially closing the significant financial gaps that currently exist. By enriching national and international policies with detailed, actionable intelligence, UCI facilitates the strategic deployment of resources, improving the efficiency and effectiveness of efforts aimed at mitigating climate change and promoting sustainable development.

Global collaboration is essential for addressing the complex challenges posed by climate change. By standardizing climate analytics across borders, UCI offers a common language that can significantly enhance the alignment of international policies among individual countries. This harmonization is crucial for ensuring that global efforts are not only coordinated but also mutually reinforcing, thereby maximizing the impact of each nation's commitments to, for example, the Paris Agreement.

UCI's decision analytics can play a transformative role in local governance. Cities around the world, such as Barcelona with its urban heat island mitigation strategies, utilize high-resolution, globally-calibrated climate data to tailor interventions that address specific local challenges. This approach allows for the micro-targeting of initiatives, such as increasing green spaces or enhancing flood defences, which are calibrated to the unique socio-economic and geographical contexts of each area. By bridging the gap between global intentions and local actions, UCI sets the stage for a more resilient and collaborative future, aligning focussed actions with the interconnectedness of Earth's natural systems.

In response to the biodiversity crisis, which poses significant threats to ecosystems and species worldwide, NbS have emerged as a critical approach. These strategies harness ecosystems and natural processes to tackle societal challenges, particularly focusing on environmental degradation and climate change impacts. UCI is uniquely positioned to elevate NbS by leveraging comprehensive environmental data to enhance biodiversity and ecosystem resilience. Integrating diverse data streams – including satellite imagery, climate forecasts, and biological diversity metrics – UCI enables precise mapping and management of natural resources. This integration aids in the strategic planning and implementation of NbS by identifying critical areas for conservation, restoration and sustainable use.

A prime example of how UCI can be applied to NbS is in the support of the 30 × 30 program, which aims to protect 30% of the planet's land and oceans by 2030. This ambitious goal requires precise planning and robust management to ensure that conservation efforts are both effective and equitable at a country level. UCI integrates diverse data streams – including satellite imagery, climate models, and biodiversity metrics – to create a holistic view of the planet's ecological health. This integration is crucial for identifying key areas that, if conserved regionally, would contribute significantly to global biodiversity and climate stability. By mapping critical habitats and assessing their current health and resilience, UCI can help policymakers and conservationists prioritize areas for protection under the 30 × 30 initiative. This targeted approach ensures that conservation efforts are focused on regions where they can have the greatest impact, such as biodiversity hotspots or key carbon sinks. Additionally, UCI's predictive capabilities allow for the modeling of future environmental and climate scenarios, providing valuable foresight into the long-term benefits and challenges of protecting specific areas.

With the capacity to overlay ecological data with human demographic and economic information, UCI ensures that the 30 × 30 conservation efforts are socially equitable. This capability is vital for designing NbS projects that not only safeguard biodiversity but also support local communities, potentially transforming protected areas into sources of sustainable development. For example, UCI could help identify regions where conservation efforts could enhance local economies through ecotourism or sustainable harvesting practices, thereby gaining community support and ensuring the success of the 30 × 30 objectives. The strategic application of UCI to the 30 × 30 program exemplifies how technology can facilitate more informed and effective conservation planning. By providing detailed, actionable insights, UCI empowers stakeholders across the public and private sectors to make well-informed decisions that align with both immediate environmental needs and long-term sustainability strategies.

Geopolitical security is increasingly influenced by the multifaceted challenges posed by climate change, with climate shifts precipitating disruptions in global stability. As a powerful tool for forecasting and mitigating the geopolitical risks associated with these climate impacts, UCI can ensure that nations are prepared for the evolving security landscape.

By leveraging advanced data analytics, UCI can help to predict how climate change can exacerbate geopolitical tensions. Analyzing trends in climate data, such as changes in precipitation patterns, temperature fluctuations, and sea-level rise, allows UCI to identify regions at higher risk of resource scarcity, critical infrastructure failure, food insecurity, migration and conflict. This predictive and scenario analysis capacity is vital for anticipating areas where climate change could destabilize governments, provoke regional conflicts or trigger mass migrations. Armed with this knowledge, global leaders and policymakers can initiate pre-emptive measures to mitigate these risks, such as diplomatic initiatives, resource allocation strategies and development aid, aimed at stabilizing potentially volatile situations before they escalate.

In the realm of defence and security, strategic planning must now account for climate-induced geopolitical shifts. UCI can support this planning by providing detailed analyses of how climate change impacts national and international security landscapes. For example, UCI could inform naval operations by predicting the impacts of sea-level rise on naval bases, or aid in preparing for humanitarian missions by forecasting areas that will be hit hardest by climate-induced natural disasters. By integrating UCI into their strategic planning, security agencies and defence departments can better prepare for the new dynamics of global security, ensuring they are ready to respond to climate-related threats efficiently and effectively.

UCI's role in enhancing geopolitical security underscores its broad applicability and vital importance in addressing modern challenges. By enabling the prediction and strategic management of climate-induced risks, UCI not only enhances national and global security but also supports the development of proactive international policies and cooperation. This comprehensive approach ensures that countries are not only prepared to defend against direct threats but are also equipped to manage the complex web of implications brought about by a changing climate.

Illustrative Case Study on Food Security: Climate-Adaptive Agricultural Practices in Agri-Intensive Regions

Context and Challenges: Agraria, a developing country characterized by its heavy reliance on agriculture, faces daunting challenges due to the impacts of climate change. Erratic rainfall patterns and an increase in pest damage significantly threaten Agraria's agricultural output, which is a cornerstone of its economy and critical for national food security. These climatic variations jeopardize the livelihoods of local farmers and the overall stability of Agraria's food supply, demanding urgent and innovative approaches to agriculture.

UCI-Powered Solutions: In response to these challenges, a non-profit organization, in collaboration with local agronomists, implemented UCI to transform Agraria's agricultural practices. UCI's advanced data analytics provided detailed insights into the region's shifting climate conditions, enabling the development of tailored, resilient agricultural strategies. These strategies included diversifying crops to reduce dependency on any single crop yield, implementing soil moisture conservation techniques to combat water scarcity and utilizing predictive analytics to pre-emptively address pest and disease outbreaks.

Impact on Food Security and Livelihoods: The introduction of UCI-informed agricultural practices marked a significant turning point for Agraria. These practices led to enhanced crop yields and increased resilience against climate shocks, thereby bolstering the nation's food security. For local farmers, the shift meant more stable incomes and reduced vulnerability to climate variability. Additionally, the promotion of climate-adaptive crops opened new market opportunities, aligning Agraria's agricultural sector with global sustainability trends and enhancing economic viability.

Expansion and Scalability: Encouraged by the success of these initiatives in pilot regions, plans are underway to expand these UCI-powered solutions across Agraria and potentially to neighboring countries. The scalability of these practices is supported by UCI's capability to adapt solutions to diverse micro-climates and socio-economic conditions, ensuring that the benefits of climate-smart agriculture can be realized on a larger scale.

Sustainable Development and Global Implications: Agraria's advancements in climate-adaptive agriculture not only contribute to the national agenda but also align with global efforts to achieve the Sustainable Development Goals (SDGs), particularly those related to combating hunger and promoting sustainable agricultural practices. The case of Agraria serves as a model for international collaboration, demonstrating how UCI can underpin widespread adoption of climate-smart agricultural practices that support both rural development and environmental sustainability.

Because of its intrinsic design, UCI retains its rigor and effectiveness down to an individual asset level. This precision supports concentrated regional strategies for localized resilience, enhancing both community-level initiatives and broader international collaborations. UCI's applications prove equally powerful whether applied at the community, city or country level, particularly in activities such as regional adaptation pooling. This approach facilitates the sharing of resources, information and strategies, optimizing adaptation efforts across geographic areas where climate impacts often transcend political boundaries. In such contexts, UCI can play a crucial role by providing precise, actionable data that significantly enhances the effectiveness of these collaborative efforts.

UCI's comprehensive environmental and climate data are instrumental in identifying regional vulnerabilities and strengths. By offering detailed insights into climate trends, risks and impacts, UCI enables regional policymakers to tailor adaptation strategies that are not only specific to their local needs but also complementary across the region. This could include coordinated responses to managing shared water resources, or developing joint infrastructure projects that are resilient to climate change.

When combined with progressive climate policies, UCI can help regions implement more effective and forward-thinking strategies. For instance, UCI data can support the establishment of regional climate standards and regulations that align with the latest scientific understanding and predictions. These policies could encourage or mandate reductions in carbon emissions, promote sustainable land use, or enforce building codes that increase resilience against extreme weather events.

As mentioned previously, financing is often a significant challenge for adaptation projects. UCI can enhance municipal financing strategies by providing quantified data that justifies investment in certain adaptations. For example, UCI can demonstrate the cost benefits of infrastructure projects like green roofs or permeable pavements in urban areas. Regions can pool financial resources to fund large-scale projects that benefit all involved parties, potentially accessing better financing rates or shared grants and subsidies designed to support multi-entity resilience projects.

The precision of UCI can also ensure that these strategies are economically viable and politically feasible, exemplifying a dynamic model of climate resilience that can be replicated globally.

Illustrative Case Study: Urban Resilience Transformation Using UCI in Metroville

Context and Background: Metroville, a developed coastal city, faces significant climate-related challenges including increased flooding, urban heat islands and energy inefficiency. In response to these challenges, Metroville has embraced UCI to enhance its strategic planning and ensure sustainable urban development.

UCI Implementation: Metroville's government implemented UCI to harness detailed, actionable insights into the city's specific climate risks and vulnerabilities. This integration into urban planning and infrastructure development has been crucial. For instance, UCI has enabled Metroville to:

Balance Investment in Adaptation and Mitigation: Understand precisely where and how much to invest in adaptation efforts versus mitigation efforts. UCI provides Metroville with granular, quantified insights, allowing the city to make unified decisions that optimize resource allocation. This helps in striking a balance between immediate adaptive measures, such as flood defenses and long-term mitigation strategies like carbon reduction.

Resilience in Mitigation Investments: Factor in resilience and adaptation even in their mitigation investments. For example, as Metroville invests in green energy solutions and enhancements to their energy grid, UCI ensures these infrastructures are resilient to future extreme weather events. This approach allows the city to not only focus on achieving a net zero goal but also ensures that its infrastructure is sustainable and resilient in the long term.

(*continued*)

Tangible Outcomes: The strategic application of UCI has led to a significant reduction in flood-related damages due to improved drainage systems informed by predictive risk maps. Initiatives to mitigate the urban heat island effect, like green roofing and increased urban forestry, were prioritized and effectively implemented, leading to cooler urban temperatures and improved air quality.

Economic and Social Impact: The economic benefits of these UCI-informed initiatives have been substantial, with cost savings from averted disaster damages and increased investment in sustainable technologies boosting the local economy. Socially, the city has seen improvements in public health and quality of life, with reduced climate-related health risks and enhanced urban environments.

Lessons Learned and Replicability: Metroville's experience underscores the importance of integrating comprehensive climate intelligence in urban planning. The city's approach serves as a replicable model for other urban areas, demonstrating how UCI can facilitate a holistic approach to climate resilience that aligns adaptation and mitigation efforts effectively.

The need for climate resilience reaches every community across the globe and is felt particularly acutely in low-income regions, which often bear the brunt of climate impacts. As such, we believe that the democratization of UCI is not just a technical requirement but a societal imperative. Integrating UCI into existing plans such as the C40 network (a global climate network of cities collaborating on climate action), and making insights accessible to everyone equips communities with the crucial data they need to drive decisions, actions and advocacy effectively. By democratizing this intelligence, communities are empowered to participate actively in their climate resilience efforts. Complex data can be translated into accessible insights to support climate literacy throughout communities, facilitating informed community engagement and inclusive decision making – ensuring no one is left behind.

UCI brings a new level of precision to discussions relating to climate action. By rigorously quantifying the impacts and embedding socio-economic data, UCI helps uncover the disproportionate effects of climate change, guiding policies to effectively mitigate disparities such as those highlighted in the most recent UN Adaptation Gap Report. This approach not only supports crafting targeted interventions but also mobilizes comprehensive support for necessary changes. Democratized UCI can also facilitate a more equitable distribution of resources, ensuring that investments are directed where they are most critically needed. It helps policymakers and planners prioritize funding and policy focus based on detailed climate vulnerability assessments, maximizing the impact of every dollar spent toward climate resilience. UCI also refines the allocation of investments by identifying the best uses for private, public, and blended capital, ensuring that financial resources are used where they can have the greatest impact.

In addition to optimizing investment, UCI plays a critical role in monitoring the effectiveness of climate interventions. By tracking the impact and calculating the returns on investments, UCI confirms the value of current strategies and informs future actions, ensuring that climate strategies remain robust and responsive to evolving needs.

By adopting UCI widely, communities and leaders worldwide can ensure their climate resilience efforts are informed, inclusive and effective. This collaborative approach to integrating UCI into climate strategies promises a more informed and equitable path toward tackling one of the most pressing challenges of our times.

The Path Forward for Climate Opportunity

UCI empowers cities, regions and entire nations to pivot from reactive to proactive stances in climate strategy. By harnessing the detailed insights provided by UCI, local governments and policymakers can devise and implement climate solutions that are finely tuned to the specific needs and vulnerabilities of their communities. This precision enables not only the crafting of policies that resonate with the immediate concerns of citizens but also the alignment of these strategies with long-term, globally-aligned sustainability goals. UCI also enables policymakers to proactively make investments that can drive strategic and competitive advantage for businesses in their regions.

The opportunity of widespread adoption and integration of UCI into global, regional and local climate strategies represents a truly transformational super-power. As digital infrastructure continues to expand and our collective scientific and technological expertise grows, the potential for UCI to drive meaningful change becomes even more pronounced. Crucially, the democratization of climate intelligence, ensuring that it is accessible to all – from policymakers to the public – is crucial for fostering a globally inclusive approach to climate resilience.

The Future of Organizational Climate Intelligence

The trajectory of climate intelligence is set against a backdrop of increasing regulatory demands, societal expectations for radical transparency and a burgeoning demand for accountability from all sectors of society. These forces are not just trends but inescapable shifts that will continue to reward or punish organizations based on their approach to climate responsibility.

Regulatory frameworks around the globe are tightening, with an ever-greater emphasis placed on comprehensive climate disclosures, financial materiality and sustainable practices. These regulations are transforming from mere guidelines to stringent mandates, requiring organizations to adopt proactive climate strategies or face significant repercussions. Simultaneously, the societal demand for radical transparency has escalated, with stakeholders from investors to consumers seeking deeper insights into the environmental impact of their investments and purchases. This is all happening while we are and will continue to experience increased frequency and intensity of extreme events across the globe. These forces together drive the need for UCI to deliver accurate, real-time data that organizations can use to inform their strategies and investments.

Looking ahead, the role of UCI in shaping governance models is poised to expand significantly. As enterprises and societies alike grapple with the complexities of climate change, UCI offers a framework for participatory decision making, integrating insights from various stakeholders to create more democratic, tailored and responsive governance structures. By leveraging UCI, businesses can not only comply with evolving regulations but also engage more effectively with their communities, aligning business practices with societal values and expectations.

The integration of Gen-AI with UCI is essential for unlocking the full potential of climate intelligence. By combining advanced AI capabilities with comprehensive climate data, organizations can gain a competitive edge, accelerate decision making, and drive innovation. This powerful combination empowers stakeholders to understand, quantify, and act on climate risks and opportunities with unprecedented speed and accuracy.

We envision a future where climate intelligence becomes a cornerstone of business and societal decision making. It is imperative for societies and enterprises to find common ground in resilient value creation. This collective movement toward climate responsibility can drive significant change, positioning UCI and Gen-AI at the forefront of our strategies to combat climate change and protect our planet for future generations. As the last generation with the power to materially alter the course of climate change, we must harness the full potential of these technologies, paving the the way for the future of autonomous, AI-powered Sustainability Agents in building a sustainable and equitable future.

Organizations that embrace this shift, integrating UCI and Gen-AI into their operations and governance models, will not only safeguard their own futures but also contribute to a more sustainable and equitable world. It is a call to action for every leader, policymaker and citizen: to unite in the pursuit of a resilient, hopeful and sustainable future.

ACKNOWLEDGMENTS

Our deepest gratitude goes to Dr. Claire Huck, whose unwavering support and tireless advocacy were instrumental in bringing this book to life. Her expertise as a climate scientist and her dedication to integrating climate considerations into everyday decisions are truly inspiring. Also, we are indebted to Dr. Ben Calderhead for his counsel on the art of computational statistics, machine learning, and deep pragmatism.

We extend our profound thanks to the incredible teams at Earthena AI and Cervest. Your passion, dedication, and courage in building a better world fuelled this project. We are grateful for your tireless work and the innovative spirit you bring to creating a more sustainable future.

Finally, we acknowledge the invaluable contributions of our friends, partners, advisors, investors and customers. Your insights and perspectives have been instrumental in shaping our thinking. We recognize that clarity often comes from unexpected places, and we are grateful for the collaborative journey we have shared.

ABOUT THE AUTHORS

Iggy Bassi is a visionary entrepreneur and a leading authority in climate intelligence. He champions innovative solutions at the intersection of climate change, artificial intelligence and resource security, driving the transition to a sustainable future for businesses and society. Iggy's expertise has been featured in top-tier publications such as *The Wall Street Journal, Financial Times*, *Axios* and *The New York Times*. He is a sought-after speaker at global events, including COP26, Cog-X, BBC News, Harvard University and the World Bank.

With his strategic acumen, Iggy helps Fortune 500 companies and governments navigate the challenges of climate change, resource security and competitiveness. He is currently the co-founder & CEO of Earthena AI, a pioneer in climate value analytics that unlocks opportunities in the new climate economy. He previously founded Cervest, an early pioneer in the emerging climate intelligence market. Earlier in his career, he worked as a strategy consultant at the Monitor Group and in mergers and acquisitions for a US investment bank, advising global tech firms including Microsoft, IBM, Telefonica and SoftBank. He also built an enterprise software business early in his career.

Iggy holds an MPhil from Cambridge University (ESRC scholarship) and serves as an advisor to several organizations. He previously authored a book on European private equity. Iggy currently lives in London.

Karan Chopra is an entrepreneur and operator at the intersection of technology and sustainability. His experience is in technologies and business models for impact at scale.

Karan's work has been featured in leading publications, including *The Wall Street Journal*, *The New York Times*, *The Guardian* and *Fortune*. He has been a featured speaker at leading forums including Aspen Ideas Festival, Harvard Business School and a storyteller for The Moth.

Karan is the co-founder of Earthena AI, empowering enterprises to drive value through climate action. He previously built leading climate intelligence and sustainable agriculture ventures with Iggy. Karan has served as advisor and board member to multiple ventures, including X (formerly Google X) on AI, robotics and the future of work. Karan is also the co-founder of Opportunity@Work, a social enterprise using large-scale labour market analytics for a more inclusive economy.

Earlier in his career, Karan was a consultant at McKinsey & Company advising governments and companies on sustainability and climate issues. He holds a BS from Georgia Tech and an MBA from Harvard Business School as a Baker Scholar. He was named a 30 under 30 Social Entrepreneur by *Forbes* in 2014 and an Aspen Institute New Voices Fellow. Karan was born in India, grew up in Ghana and currently lives in the San Francisco Bay area.

INDEX

Page numbers followed by *f* refer to figures.

INDEX

INDEX